# *Figuratively Speaking*

## In the Computer Age

### Techniques for Preparing and Delivering Presentations

**Authors**

David M. Orchard

Barry D. Perrow

Kenneth Frakes

Torin B. Reed

With a special introductory chapter by Gene Shinn

Printed in the U.S.A.

ISBN: 0-89181-822-7



Association Editor: Neil F. Hurley
Publications Manager: Kenneth M. Wolgemuth
Managing Editor, Publications: Anne H. Thomas
Design: Carl Brune
Cover Illustration: Rusty Johnson
Composition: Manzanita Graphics L.L.C., Denver, Colorado



This and other AAPG publications are available from:

The AAPG Bookstore
P.O. Box 979
Tulsa, OK 74101-0979
Telephone: 1-918- 584-2555 or 1-800-364-AAPG (USA)
Fax: 1-918-560-2652 or 1-800- 898-2274 (USA)
www.aapg.org

Australian Mineral Foundation
AMF Bookshop
63 Conyngham Street
Glenside, South Australia 5065
Australia
Tel. +61-8-8379-0444
Fax +61-8-8379-4634
www.amf.com.au/amf

Geological Society Publishing House
Unit 7, Brassmill Enterprise Centre
Brassmill Lane, Bath, U.K.
BA1 3JN
Tel +44-1225-445046
Fax +44-1225-442836
www.geolsoc.org.uk

Affiliated East-West Press Private Ltd.
G-1/16 Ansari Road Darya Ganj
New Delhi 110 002
India
Tel +91 11 3279113
Fax +91 11 3260538
e-mail: affiliat@nda.vsnl.com

# *Authors' Biographies*

David M. Orchard is president of Manzanita Alliances, Inc. in Houston, Texas, a firm that provides graphics and writing services to corporate clients. Mr. Orchard has a background in petroleum geology, has published several technical papers, and has delivered a large number of professional presentations.

Barry D. Perow is president of Manzanita Graphics, LLC in Denver Colorado. His company specializes in presentation graphics, publishing, and electronic media for the oil and gas industry. Mr. Perow was a contributing author to the original *Figuratively Speaking*.

Ken Frakes is a communications management consultant in Tulsa, Oklahoma. Formerly, he was publications production manager for AAPG, executive director of the Printing Industries Association of Oklahoma/SW Missouri, and chief operating officer of a printing firm.

Torin B. Reed is managing partner of Aardvark Imaging, LLP in Denver, Colorado. His company is an output service bureau providing large format and presentation products for a variety of industries. Mr. Reed's background is in traditional color photo production, and he made the shift to the digital world several years ago.

# Contents

# *Acknowledgments*

We wish to acknowledge the editor and authors of the original *Figuratively Speaking* for their contributions to the art of presentation graphics and for setting the standard for books on this subject: Duncan Heron, Eugene A. Shinn, Leon H. Gellman, Barry D. Perow, Lev Ropes, Jerome P. O'Neill, Jr., and Robert L. Freed.

Several graphic artists and desktop publishers at Manzanita Alliances, Inc., and Manzanita Graphics, L.L.C., contributed to the concepts, illustrations, and design of this book, including Chris Chavez, Darin Dickson, Donna Harwell, Ruth Klein, David Lemm, Anthony Mata, and Antoinette Williams.

We thank Kenneth Aitkens, Barbara L. Faulkner, Hugh Hay-Roe, and Alan J. Scott for their helpful reviews and critical comments.

Rusty Johnson of the American Association of Petroleum Geologists drew the cover art and the various pictures of geese. Tom Hendricks contributed the photographs in Chapter 5.

Software cited in this book include PowerPoint®, a registered trademark of Microsoft Corporation; Illustrator®, PageMaker®, and Photoshop®, registered trademarks of Adobe Systems, Inc.; Lotus 1-2-3®, a registered trademark of Lotus Development Corporation; Canvas®, a registered trademark of Deneba Software; QuarkXpress®, a registered trademark of Quark, Inc.; WordPerfect® and CorelDraw®, registered trademarks of Corel Corporation, AutoCAD®, a registered trademark of AutoDesk, Inc.; WinZip®; a registered trademark of Niko Mak Computing, Inc.; Stuff-itTM, a registered trademark of Aladdin Systems, Inc.; and FreeHand® and Director®, registered trademarks of Macromedia, Inc. Macintosh® is a registered trademark of Apple Computer, Inc. SyQuest®, Zip®, and Jaz® are registered trademarks of Iomega Corporation. Kroy™ is a trademark of Kroy LLC.

Many of the illustrations have been redrawn from figures published in professional journals. We acknowledge the source of these figures on page xi.

# *Preface*

Oral presentations have long been valued as a means of communicating scientific and technical information to professional audiences. These presentations take place in many different venues, from classrooms to convention centers, in informal gatherings and at national meetings of professional societies. Almost all make use of visual aids—slides, viewgraphs, posters, and, increasingly, projections directly from computers using digital projectors—to illustrate the subject matter.

In 1986, the American Association of Petroleum Geologists (AAPG) published *Figuratively Speaking* to address the design and production of 35-mm slides for effective presentations. That popular book came out immediately before computers became the standard tools for graphic artists and featured techniques for manual and mechanical art preparation and specialized photographic methods. Since that time, we have entered the computer age in all aspects of technical communications, including presentation graphics. In 1986, the language of a drafting department included Kroy, Rapidograph, airbrushing, tape, t-squares, and drafting tables. Now, the jargon of graphics is an alphabet soup of TIFF, BMP, JPEG, DXF, and CGM. The tools and techniques now available to us greatly enhance our ability to effectively illustrate technical information and incorporate those illustrations into presentations.

The original *Figuratively Speaking* was often referred to as AAPG's "slide manual." Today, 35-mm slides remain a common means of illustrating oral presentations. However, the graphics world is rapidly changing as speakers turn to electronic presentations, making use of digital projectors, Web sites, and other new media. In recognition of this, we could replace the word "slide" with "visual" or some other such term. Or, we could expand the sense of "slide" beyond that of a projectable, photographic product to include electronic displays. In this book, we have chosen to do the latter.

Much has changed since the publication of *Figuratively Speaking* and, as a result, much has changed in this rewrite of the book. Major emphasis remains on the principles of good slide design that are universal, regardless of how the art is produced. We have retained and expanded discussions on preparing for professional presentations, and we have extensively rewritten the discussion of slide design. New topics include electronic presentations, poster sessions, working with service bureaus, and a glossary of computer graphics terminology. In addition, we have expanded the subject matter of the illustrations beyond geology to include examples from several other technical disciplines.

This book is not a software manual. Much too much applicable software exists to be described in this short treatment, and that software is changing and advancing extremely rapidly. Further, it is difficult to illustrate in book form many of the exciting features of computer graphics, such as animation, embedded video, and special effects. Our purpose, therefore, is to define standards for presentation graphics, to add to your lexicon of computer-related terms, and to describe the capabilities available to you through computer graphics. You are on your own to choose and learn the software that best suits your purpose.

This book discusses technical presentations, from preparation to delivery. Our hope is that this advice will allow you to shine in front of your audience, help your peers receive the information you intend to convey, and serve the process of technical communication.

# *Figure Credits*

| | |
|---|---|
| 3-1 and 5-5 | Redrawn from Department of Energy, 1992. *National Energy Strategy: Powerful Ideas for America*. DOE/S0083P, Washington D.C. |
| 3-4 | Redrawn from Mpodozis, C., and K.S. Mahlburg, 1992. Upper Paleozoic to Triassic evolution of the Gondwana margin: Evidence from the Chilean Frontal Cordilleran batholiths. *GSA Bulletin*, v. 104, #8, pp. 999-1014. |
| 3-2, 5-5 | Redrawn from Energy Information Administration, 1992. *Electric Power Annual*, 1992. Department of Energy, Washington D.C. |
| 4-5, 5-30 | Redrawn from O'Hare, G., 1997. The Indian monsoon: Part 2 The rains. *Geography*, v. 82, pp. 335-352. |
| 5-32 | Redrawn from Marwick, M., 1997. The population of the Maltese Islands, Census 1995. *Geography*, v. 82, pp. 179-182. |
| 5-3, 5-4 | Redrawn from Fitzpatrick, E.A., 1980. *Soils: Their Formation, Classification, and Distribution*. Longman Group Ltd. |
| 5-8 | Redrawn from Neu, Harold C., 1992. The crisis in antibiotic resistance. *Science*, vol. 257, pp. 1064-1073. |

# 1

# *A Goose is Loose in the Audience*

Gene Shinn has waged a long crusade for improvement in the quality of presentations at professional meetings. As you can read in this article, he has experienced the frustration that comes from sitting through poorly planned, poorly illustrated, and poorly delivered talks.

A slightly modified repeat from the original *Figuratively Speaking*, this article speaks directly to the issue of ensuring that both speakers and audiences receive the greatest possible benefit from presentations.

First there was a snicker, followed by a cough or two, then a crescendo of "honk, honk, honk," as the unmistakable calls of geese came from the rear of a dimly lighted conference room. This was the annual meeting of a major geological society.

Time? Sometime soon.

Why? Well, let's get on with the story and see....

The program committee had been struggling with the problem for years. Attendance was dropping, costs were climbing, and without attendees, sponsors and exhibitors were cancelling. The society was in debt, and there was fear that annual conferences would soon fade into the past. Something needed to be done quickly. The quality of presentations needed to be improved.

For weeks, the committee struggled. Two philosophical camps evolved within the committee; division revolved around a philosophical question of which is more important, quality of information or quality of presentation. Lined up on one side were academics and researchers, while on the other were the convention organizers and attendees. A deadlock continued until a drilling company put up money for 5,000 goose calls. The plan was to equip each attendee with a goose call on a string, to be worn around the neck. Admittance to session rooms required a goose call. The instructions advised, "If you cannot understand a presentation because slides are unreadable or confusing, let the speaker know!"

Traditional egg, cabbage, and tomato throwing had been proposed, but the committee considered that too messy and, besides, the supply of eggs would have run out during the first morning session. A "worst presentation" award also was considered, but they feared the competition might be too keen and the society could only afford a few plaques.

Ridiculous? Of course, but who can deny the eloquent effectiveness of such drastic measures?

Various guides for making slides and techniques for giving presentations have been available (and largely unused) for years. We are all educated people, right? One would think that after four to eight years of college, such smart people would have learned how to prepare slides and give oral presentations. The truth is that most of us have learned little about the art of presentation. Very few geology students have the opportunity to take a course designed to improve oral presentations. Even if available, such courses usually are not required, so we avoid them.

Then why not learn by osmosis, simply by exposure to lecturing professors? The truth is that they usually don't know how to give an effective 15-minute lecture. What we learn about presentation techniques from most professors may be their bad habits. You see, they learned the same way we did, from their professors, and the cycle is perpetuated.

Recall that students are captive audiences. Are students going to tell a professor that he or she doesn't know how to teach?

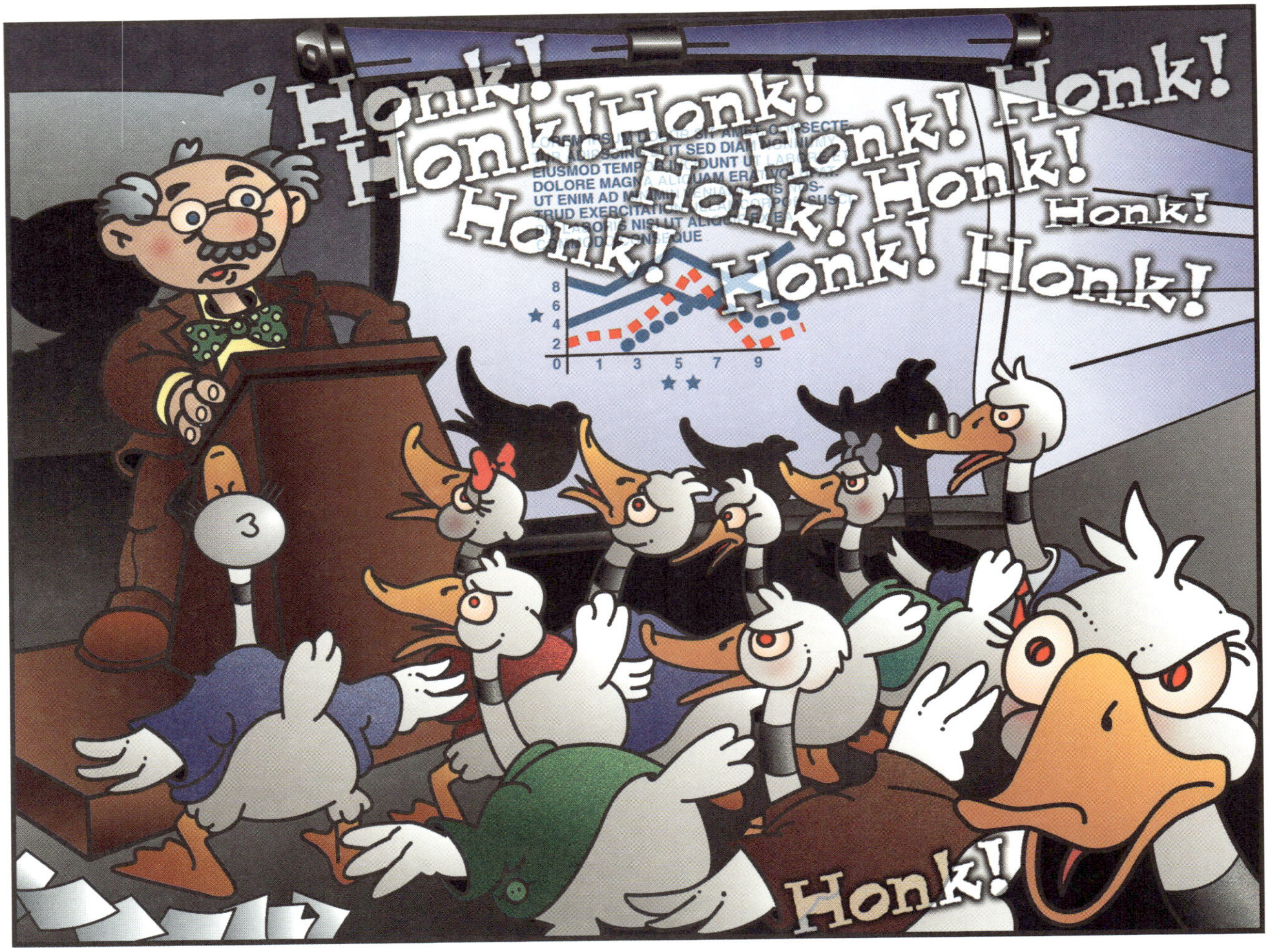


Besides, classroom teaching is generally done using the cornerstone of education, the chalkboard rather than the ruler. Chalkboards are great, they are cheap, they are easy to see in a classroom, and students can ask questions if something is unclear. Chalkboards also give many teachers tremendous power. Write something on the board and a roomful of students unquestioningly copies it as if the words came from the mountain on stone tablets. (Now that I have alienated my friends in the teaching profession, I might as well stick my foot in farther!)

Teachers are role models for nearly all of us, and they hold our respect for years; but how good are they as role models when it comes to giving presentations? You see, teachers put in so many hours with captive audiences, the quality of their presentations is rarely considered. Furthermore, a professor has an entire semester to deliver all pertinent material and can use hour-long class sessions to elaborate and develop concepts using all supporting data.

When slated to give a short oral presentation, the professor who is used to lecturing in one-hour segments is suddenly faced with condensing months of research into a 12– to

15–minute talk. But habit dictates the presentation of all supporting data, including lists of formation names, faunal names, and stratigraphic sections. Frequently, to save time and expense, these are photographed from an existing document and justified with comments such as: "I apologize for the slide," "I know you can't see these numbers down here...," and "Oh, I see I'm running out of time!"

At times like these, I wish I had a goose call!

This often-repeated scenario is usually an honest mistake, but sometimes it almost seems deliberate. Surely we have all at some time left a meeting scratching our heads and saying, "I must really be dumb. I never did get what he was driving at." Maybe, just maybe, that was the speaker's intent. Or maybe the speaker didn't have a point! Too bad!

So, these are our role models, and when those of us who get the call go into teaching, we recycle the carefully learned bad habits once again. Possibly what contributes to recycling these habits is a snobbish social attitude toward those who do communicate clearly. Good speakers are sometimes considered actors, snake-oil merchants, promoters, or worse—"oil men." Funny thing, though, weeks after hearing a "snake-oil merchant's" pitch, you generally remember what was said and, believe it or not, that's what these professional meetings are all about. Think about why you remember someone's talk. Was it because it was simply stated? Was there only one main point? (Three or four at most?) Were you told what the main point would be at the onset, and was that point reinforced during and at the conclusion of the talk? Was everything subservient to proving and explaining the main point? Were the slides simple, and did they contain no more than six lines of words? Was all the available space used so as to make the letters as large as possible? This could go on and on, but it won't because how to do it right is what the following chapters are all about, and they have been prepared by people who know a lot about the subject.

Obviously, much of what I have said so far has been in a facetious manner to get your attention. Yet who among us will deny that there is a kernel of truth hidden in the satire? I know for certain that every reader out there has attended lectures containing long lists of species or other data too copious to be read in the allotted time, even if the type was large enough to read (and in such lists it seldom is). Is there anyone who has not heard a speaker break the cardinal rule by saying, "I apologize for the slides?"

To the young who are just beginning a career in science, it must sound as if I am advocating presentation quality over research quality. Certainly, I am not. Nevertheless, it is a problem to define the proper balance. A sensitive person will probably feel uncomfortable with the problem. We worry because we know there are those who can put on beautiful presentations and not tell us anything new or worthwhile. On the other hand, we

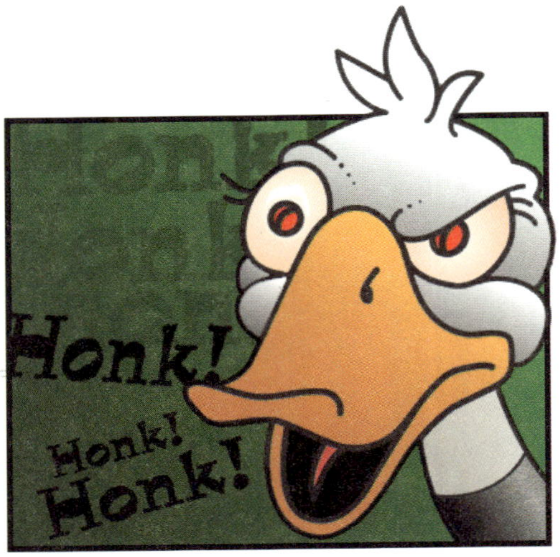


solicit money to do research and if the research results are not conveyed to others, then what has Science to show for its efforts? Why should anyone be given money to learn something if, once learned, the knowledge is not communicated to others?

While working for a major oil company research laboratory, I was prodded into preparing monthly reports. Naturally, I complained bitterly that reports were unnecessary and that the time spent preparing them was better spent doing the research. Management countered by pointing out that those reports were the only tangible product of the money spent. Those reports became even more important when it was pointed out that they were worth about $30,000 per page! So, how much is a 20-minute oral presentation worth?

Now with all of the advice in this book, how can you go wrong? I'll tell you how. Refuse to learn and apply some of these techniques and you may hear the unmistakable sound of geese at your next presentation!

# 2

# *Making a Professional Presentation*

Oral presentations are a time-honored means of presenting information to colleagues, customers, and other audiences. For some, these are comfortable, effective experiences. For others, they are far less enjoyable. For all, a few basic principles of purpose, preparation, practice, and presentation are very helpful.

Professionals give electronic or slide presentations for many reasons. Some speak to convey scientific information, perhaps in front of large audiences of their peers at national or international conventions. Some speak to sell ideas or products. Some to motivate or teach. Talks take place in modern conference centers with state-of-the-art projection equipment, and they take place in hotel rooms and musty college auditoriums. For all purposes and in all venues, the speaker wants to put on a good show.

Putting on a good show is important (and that, of course, is what this book is all about!), although the effort to do so can be stressful. Speakers are expected to present concise, well-organized, well-illustrated, and, possibly, somewhat entertaining presentations, keeping their audience alert and attentive, all within strict time constraints. Even an experienced speaker who has just delivered a near-perfect talk may afterwards reach into a shirt pocket for a business card, only to find that the card is damp with sweat!

So, how do you put on a good show? There are no secrets here. Being prepared, having well-designed slides, avoiding such foibles as having your slides out of order, and delivering the talk with confidence and poise are the ingredients of success. Do well, and you can have one of the most satisfying of all professional experiences—shining in front of an audience of your peers.

Let's approach the topic in terms of:

- Purpose
- Preparation
- Practice
- Presentation

"Why give a paper anyway? It is a remarkably inefficient method of disseminating scientific information. A speaker can reach only a handful of potentially interested colleagues, and the time constraints at most meetings rarely permit an extensive exploration of a subject. A talk is an ephemeral event, of which the only lasting elements lie in a brief abstract and the fading memories of the relatively few who attended the session.

Yet oral presentations are a time-honored means of transmitting the fruit of scientific research, and scientists will no doubt be speaking at professional meetings for a long time to come. It is a way to reveal quickly our most recent thoughts and discoveries to our colleagues–a way of informing the world that we are into a specific aspect of research (staking out a bit of scientific territory, if you will). It is too often a prerequisite for obtaining funds to attend these meetings, but we won't go into that. A scientific talk unquestionably is highly effective for developing fruitful discussion with others interested in the subject. So we will continue to give talks; the question is, how do we ensure that we give good talks?"

*H. Edward Clifton, 1978*

## Purpose

More than likely, you know what you want to achieve before you first commit yourself to making a presentation. Perhaps you have already written an abstract or a paper on the subject. Perhaps you are an invited speaker, and the topic has been assigned. Perhaps you are a graduate student making the transition to the ranks of professionals and your purpose is both to communicate your work and to create a favorable impression. In any case, carefully reflecting on your purpose at the onset of preparation is very important.

"Make your conclusion slide first" (Shinn, 1981) is a bit of advice that makes perfect sense for clarifying the information that you hope to communicate. This will help define your purpose and will allow you to work backwards to build a talk that leads toward those conclusions, avoiding time-consuming digressions and the presentation of unnecessary detail.

"...few speakers do everything wrong; even fewer do everything right."

*H. Edward Clifton, 1978*

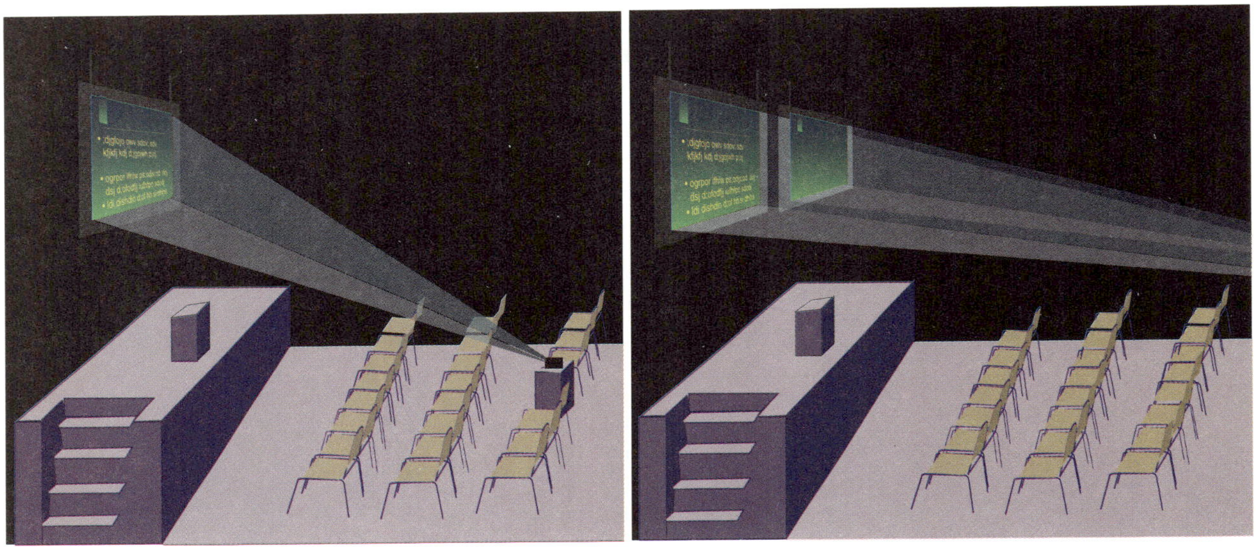

figure 2–1 *Venues may have one screen or two.*

### Questions to Ask About the Venue

- Will there be one screen or two?
- Will projection be from the front or the rear of the screens? (Remember to orient slides in the tray correctly.)
- Will there be a projectionist?
- Is a pointer supplied?
- How large will the audience be?
- Is there equipment for an electronic presentation?
- Will there be time for practice?
- Is there a separate room for practice?
- Are the screens vertical, horizontal, or square? (You may be able to use vertically oriented slides on vertical or square screens.)

With a purpose in mind, you must then consider the limitations that are imposed by the audience and the venue. If a purpose is not achievable, it must be redefined. Ask yourself these questions:

- Who is my audience, and what kind of delivery will give them the most value? Do they want a lot of details, or are they more interested in analyses or conclusions? Do they have prior knowledge, or do I have to carefully set the scene with background information? Do they all speak English?
- What limitations are imposed by the room, the projection equipment, and the time limit? What can I possibly say in the time allowed? Will I be projecting onto one screen or two (Figure 2–1)? Is time allowed for questions and answers?

Answers to these questions may come from your prior experience, from speaking to your colleagues, from session chairmen, or from an organizing committee. Ask even the most elementary questions, and you won't get off on the wrong foot and risk an unpleasant surprise.

## *Preparation*

Leaving off the most fundamental admonition about preparation—that being, of course, do not wait until the last moment—we suggest that you take the following steps:

- Write an outline (which should not digress too sharply from a previously written abstract, unless of course you have discovered that the abstract is not entirely correct.)
- List the figures that correspond to each part of the outline. Include a title slide, word slides to let your audience know where you are going and to keep you on track, and data slides, as appropriate. Don't forget your conclusion slide (the one you made first). The storyboard technique is helpful for this step (Figure 2–2).

figure 2–2 *An example of a storyboard. Some presenters prefer to first compile sketches of slides on cards or sheets of paper as a "pre-storyboard" step and add to or shuffle these as plans progress. Once slides have been designed, many presentation software programs will produce a printed storyboard. This can be useful for sending to coworkers for review and for last-minute review before a presentation.*

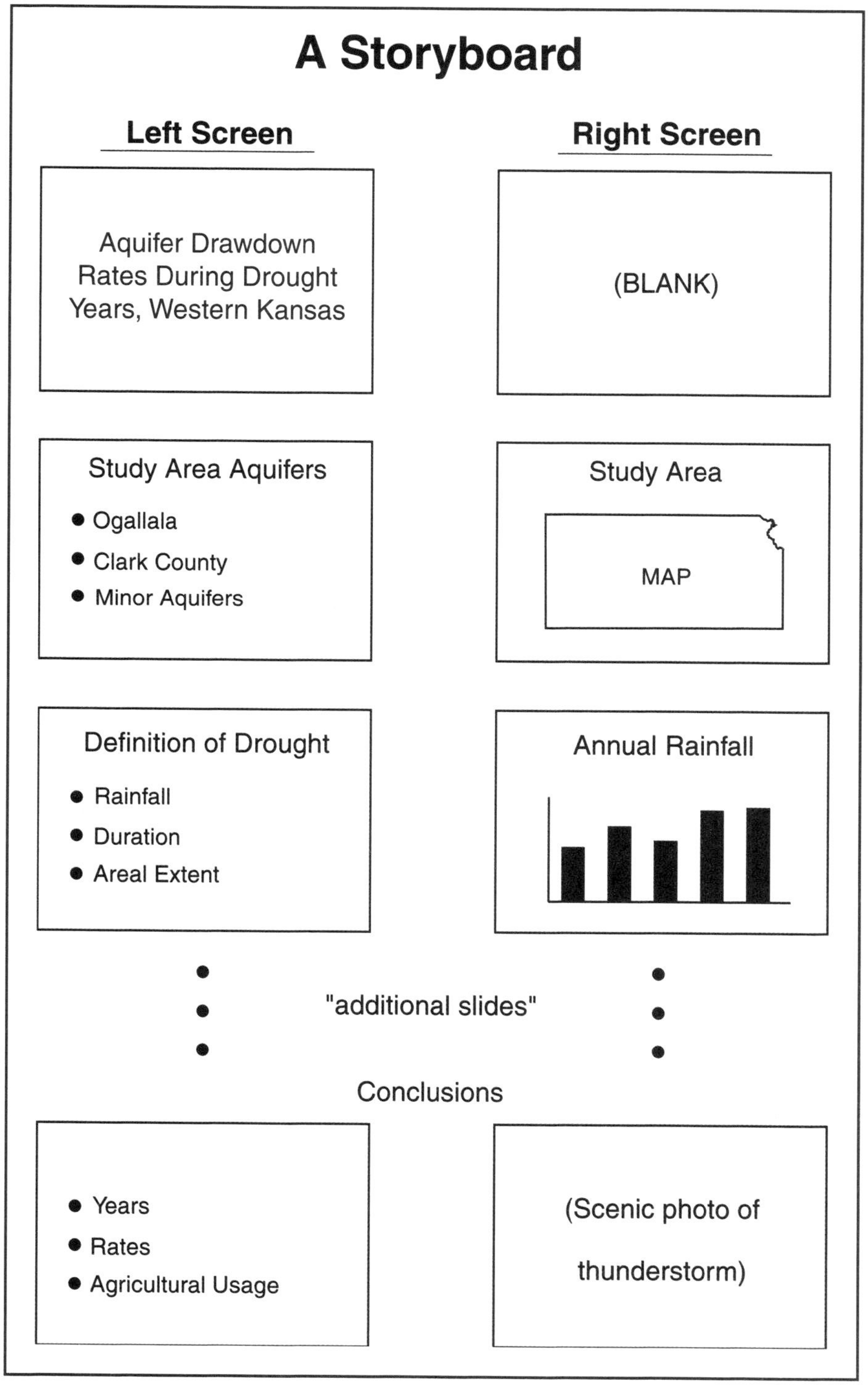

## Lost and Found

Several years ago a young geoscience professional gave a talk at a sectional meeting of the AAPG. His topic was exciting, his work was thorough, and his slides (drafted by a major oil company graphics department) were excellent. He was ready! The venue had two side-by-side screens, with the lectern stood towards the audience from the screens.

Off he went into the talk with great confidence and poise. The slides flashed onto the screens, and things were going perfectly.

Suddenly, he was lost! The slides on the two screens did not match. Poise went to pot as he started pushing buttons forward and backward to find the right slides. The audience was deathly still. Because of the location of the lectern forward of the screens, he was forced to crane his neck to watch the slides, and in so doing his head was close to the microphone. For two seemingly endless minutes the audience listened to protracted, frustrated moans as the sought-after slides refused to appear.

And then, just as suddenly, he found his place and continued in a chipper manner just as if nothing had happened!

## How many slides per talk?

- The simpler the slides, the more you can show
- One major message per slide
- Enough slides to illustrate all major messages
- Few enough to get through in time
- As many as your practice sessions indicate are best

- Make preliminary sketches of the figures (pencil sketches, cut-and- paste mock-ups, or early drafted versions).
- Expand your outline with words that correspond to each figure. What will you say when each figure is on the screen? Think of verbal transitions between slides.
- Practice your talk using your storyboard and preliminary sketches. Revise both the words and the figures, as necessary.
- Draft the figures once you are fairly sure that they won't need extensive and expensive revisions. Follow the design principles discussed in the next chapters.
- Prepare any notes that you want to carry to the lectern.

These steps leave plenty of room for iterative revisions. You can rearrange the figures at will. You can discard or add content. You can start imagining yourself in front of the audience delivering succinct statements illustrated by superb graphics, winning a hearty round of applause!

There is no substitute for the standard introduction-discussion-conclusion sequence of delivery. The adage, "Tell them what you are going to tell them, tell them, and tell them what you told them" is appropriate for professional presentations.

In all of your preparation, keep time constraints foremost in mind. Every venue imposes some type of time limitation, and the rules are often strict. If, for example, a talk is part of a moderated session, one speaker who overruns the time limit will cause problems for the other speakers and the audience.

A rule of thumb for time allocation is:

- 10% for an introduction
- 80% for presentation of methods and data
- 10% for conclusions

Allocate time for a question and answer session, if appropriate.

Pace your speech. Allow time for brief pauses before and after you introduce or describe a slide. Give the audience a chance to absorb as much information as possible from your visual aids.

Anticipate running out of time (in spite of all of this careful preparation and practice). What will you do if the previous speaker runs over and cuts into your time? How about a delay caused by an equipment or power failure? Being prepared to compress your presentation by leaving out the least critical discussions or shortening the introduction and/or conclusion will help you retain your composure, communicate as much information as possible, and finish on time.

In all of this, don't lose sight of your purpose, and don't lose sight of the limitations of the venue. Know what is achievable, and then prepare to achieve it with perfection. Get psyched!

## Practice

Several rounds of practice are indispensable for ensuring a good talk. Practice allows you to refine the sequence and the pace of the presentation, to receive feedback from colleagues, to reduce or eliminate a reliance on notes, and to ensure that you get through the talk in the time allowed.

As we have suggested, practice starts very early in preparation, when you first start compiling the figures and deciding what to say. Once the slides are ready, practice progresses to:

- Practicing aloud by yourself
- Practicing in front of an invited group of reviewers
- Rehearsing (if it can be arranged) in the actual venue

Sessions in front of an invited group of colleagues are very helpful. Ask them to be brutally honest, and take their suggestions seriously. Ask for their comments on your content, visuals, voice, and mannerisms. Arrange these sessions early enough to have time to incorporate their valid comments.

Practicing at the venue itself will go a long way towards building confidence and ensuring that things go smoothly. You may be given an opportunity to go through your slides in the actual room, possibly with the projectionist present. This will allow you to become familiar with the layout of the room, with the lectern, the pointer, and the projection equipment, and to make a final check that your slides are in order. Even if access to the room is not available, some conventions set up a separate speakers' preparation room to meet these purposes.

Time every practice session. If you can't get through in the allotted time in practice, you can't expect to in the real thing. Shorten your presentation as necessary to ensure that you will finish on time.

**Organizing Slides for One Screen or Two**

One Screen

- Organize sequentially
- Repeat reference slides; e.g. outline, reference maps, etc.

Two Screens

- Use one screen (usually the screen on the audience's left) for reference slides and the other screen for data slides
- Use side-by-side images for wider views or similar subjects
- Use blank, black slides to dim an unused screen and to avoid inappropriate pairs
- Repeat reference slides, as with one screen

## Presentation

Up to the actual start of your talk, you can refer to this book or any number of others. You can ask friends for advice. You can ask your spouse for reassurance. When the talk starts, however, you are on your own! But, of course, you are ready.

You have nothing truly to fear—there aren't actually any goose calls in the audience, and you won't read a scathing review of your talk in the following day's *New York Times*. You have thoroughly prepared and practiced your talk. Absent a power failure or a failure of nerve, things are going to go fine! Enjoy it, and let the audience see that you are pleased to share your work with them.

Deliver your talk just as you have practiced it. Your practice sessions in front of brutally honest (at your insistence) reviewers will have ensured that the talk is good. Your well-designed slides will both illustrate your points and keep you on track. You know that you can finish on time. You don't need to ad lib. Be friendly with the audience, but get on with the business of the talk.

### Credit Where Credit is Due!

You may not be speaking at the Academy Awards, but don't forget to acknowledge those who gave you intellectual, logistical, or financial support. You may owe an acknowledgment to individuals, institutions, governments, or corporations who contributed key ideas, financial assistance, or access to data.

### Loading Slides in Carousels

If your visuals are 35-mm slides, avoid the embarrassment of upside down or backward slides by loading the carousel yourself. If you trust someone else to do this for you, you are likely to have a mistake.

Many speakers mark a corner of the slide frames with a colored pen. When the slides are in the carousel, the mark will be in the same orientation for all slides, and the slides will project correctly. For example, if you mark the lower left corner of a frame, the mark will be in the upper right corner when the slide is placed upside down in the carousel for projection.

Use a carousel that holds 80, rather than 140, slides. This will help prevent "sticky" slides that don't drop.

Do not read the text of your slides, except for special emphasis. If your slides are designed correctly with a minimum of text in short, succinct phrases, the audience can read much faster than you can speak. Read the text, and they will be way ahead of you. Instead, elaborate and build on the information presented by the slide. Give the what, why, and wherefore of the information, and let the audience both look and listen.

What happens if something does go wrong? The projection bulb burns out or there is some other technical glitch? A waiter drops a tray of drinks?

In some cases, you have help. The projectionist can deal with the technical problems, and the moderator can tell you if extra time is available to make up for the interruption.

But even more important is to keep your composure and poise, and don't forget your sense of humor. After all, most everyone in your audience has stood exactly in your shoes before and will do so again. Nothing can happen that is truly the end of the world. When the interruption is over, get back on track as if nothing happened.

Most importantly, save getting visibly (and audibly) irritated or mad until you are safely away from the audience. You don't want attention to turn from an unfortunate interruption to an unfortunate outburst.

What happens if you are unlucky enough to encounter a rude or inattentive audience? Again, you may have help if a moderator is present to quiet the audience. Remember that somebody, or more likely several somebodies, in that audience is really there to hear what you have to say. Speak to them and don't worry about the rest. A few people may be noisy, but at least the geese aren't honking!

Optical or laser light pointers are popular visual aids. Speakers use them to point to features on slides while standing at a lectern well away from the screen. Used properly with a steady hand, pointers help an audience focus on important features and to follow the speaker's discussion. Used improperly with an unsteady hand, the light dances and gyrates wildly across the screen, distracting and annoying the audience.

To properly use a light pointer:

- Turn it on
- Point briefly to a feature
- Turn it off
- Repeat when and if necessary

You may also enlist a colleague to handle the laser light pointer for you from the front row of the audience. If the two of you have practiced, your colleague will be able to point to appropriate features on your slides from a position facing the screen with the pointer braced for steady projection.

With some electronic presentation programs, you can eliminate the need for a laser light pointer by using a curser, blinking highlights, or other programmed special effects that draw the audience's attention to a specific feature on a slide.

## *Does the Moderator Know You?*

If you will be introduced by an emcee or a moderator, make sure that you provide the right information for the best introduction possible. A good introduction can immediately establish credibility and audience support. Consider sending a short list of information about yourself to the moderator in advance so that he or she can prepare the remarks. Be sure to include the title of your talk.

## *Recommended Reading*


Clifton, H. E., 1978, "Tips on Talks." *Journal of Sedimentary Petrology*, Vol. 48.1, pp. 1–5.

Shinn, E. A., 1981, "Make the Last Slide First." *Journal of Sedimentary Petrology*, Vol. 51.1, pp. 1–6.

Moriarity, S., and T. Duncan, 1995, *Creating and Delivering Winning Advertising and Marketing Presentations*, NTC Business Books, Lincolnwood, IL.

Joss, M. W., 1999, *Looking Good in Presentations*, 3rd ed., The Coriolis Group, Scottsdale, AZ.

# 3

# *Elements of Technical Graphics*

Before we reach the discussions of how to design and draft slides and posters, it is useful to spend some time discussing the content of technical illustrations. The ideas presented here allow an analytical approach to determining what *can* and what *must* be shown on an illustration to best communicate the intended information.

Throughout our discussion of slide design, we emphasize that 35-mm slides are a very restrictive format for communicating technical information. They are limited by size and by readability constraints. Therefore, simplicity, clarity, and legibility are critical.

Because of these constraints, the question, "What can I leave out?" is as important in slide design as, "What must I show?" A careful choice of content is the quickest path to achieving simple, clear, and legible slides.

To answer the questions of what to leave in and what to leave out, it is useful to consider some theory. We present two related techniques (the "parts" of an illustration and the "layers" of an illustration) of understanding the content of illustrations.

## *The Parts of an Illustration*

Illustrations can be understood as comprising a series of parts, namely:

- Framework
- Data
- Interpretation
- Annotation
- Embellishment.

These vary widely depending on the type of illustration involved, and not all illustrations have all parts. Table 3–1 summarizes the characteristics of these parts.

table 3–1: *Definitions and examples of the "parts" of an illustration.*

| PART | DEFINITION | EXAMPLES |
|---|---|---|
| Framework | The basic construction of an illustration, onto which data, interpretation, annotation, and embellishment are added | • The pie of a pie chart<br>• The axes of a graph<br>• The outline of a map<br>• The cells of a table |
| Data | Actual points of data, measurable or stated on the illustration | • Bar heights on a bar chart<br>• Data points on a scatter diagram<br>• Precise drawings or photos<br>• Numbers in a table |
| Interpretation | Any depiction of interpretation, extrapolation, or conclusion | • Best fit lines on a scatter diagram<br>• Contour lines<br>• Trend lines or extrapolations |
| Annotation | Written and or numerical information that explains the framework, data, or interpretation of an illustration | • Captions<br>• Legends<br>• Labels<br>• Titles<br>• Words on a word slide |
| Embellishment | Anything that enhances the attractiveness, cleverness, or salesmanship of an illustration | • Color,shading, stippling<br>• Cartooned features<br>• Drop shadows or third-dimensions<br>• Superfluous artistry |

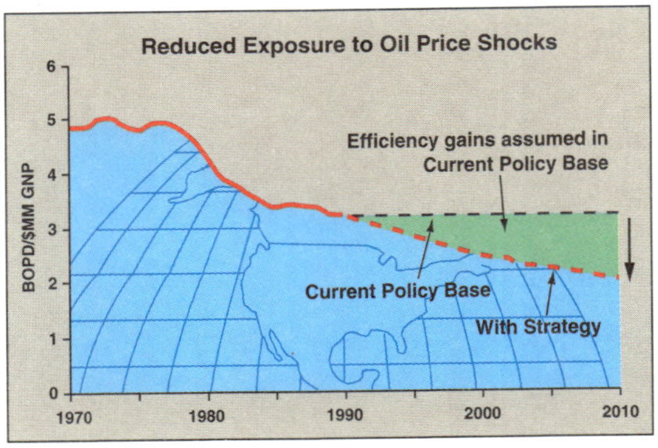


figure 3–1a *Composite Illustration*

Let's consider a few examples. Figure 3–1a is a line graph that appeared in a 1991–1992 government publication touting the National Energy Strategy. Figures 3–1b–f break this illustration into its component parts.

The framework of the illustration comprises the axes of the graph. Data are depicted in the solid red line for the years 1970 to 1990 (the years prior to publication). Diverging dashed lines from 1990 to 2010 represent interpretation (in this case, with the strongly political overtone of "follow our policy or stew in a sulfuric atmosphere!"). Annotation includes the caption, labels on the axes, and identifications of the interpreted lines. The authors added considerable embellishment, specifically the color, the superimposed map, and the down arrow emphasizing the advantages of the proposed policy.

The second example (Figure 3–2), also from a government publication, is a bar chart that depicts electrical generation by fuel source for the year 1990.

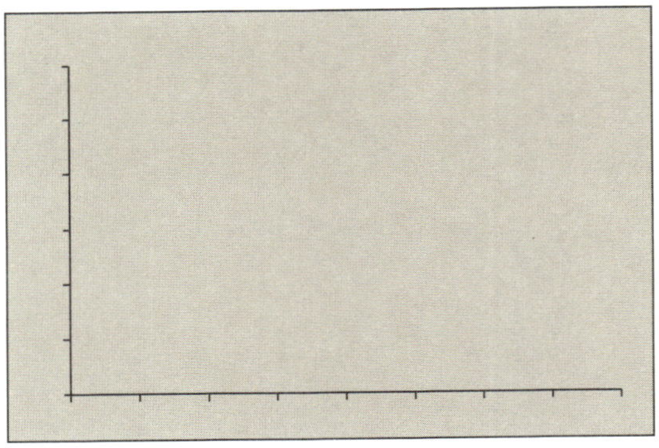

figure 3–1b *Framework*

figure 3–1 *A complex illustration broken down into its component parts.*

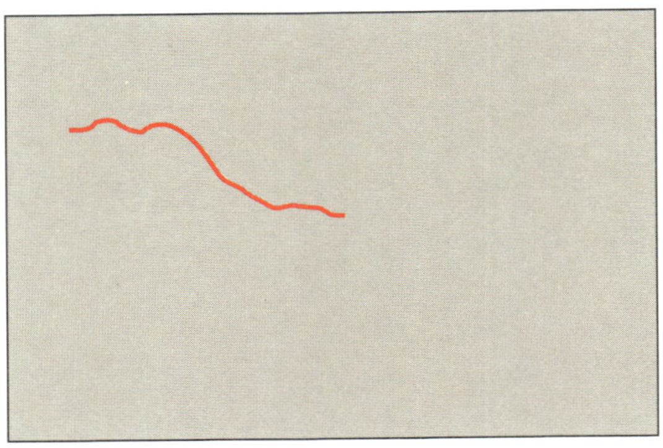

figure 3–1c *Data*

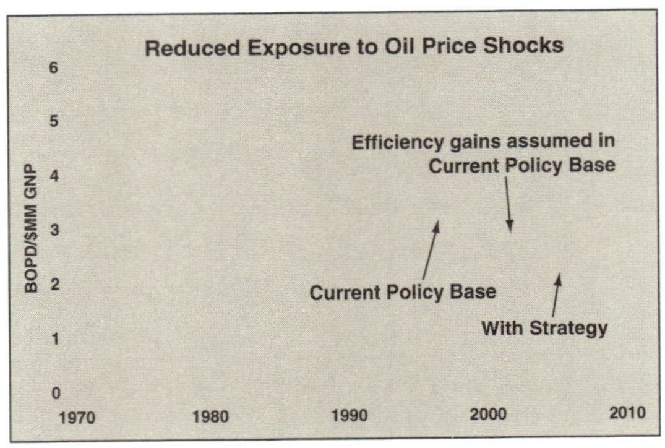


figure 3–1e *Annotation*

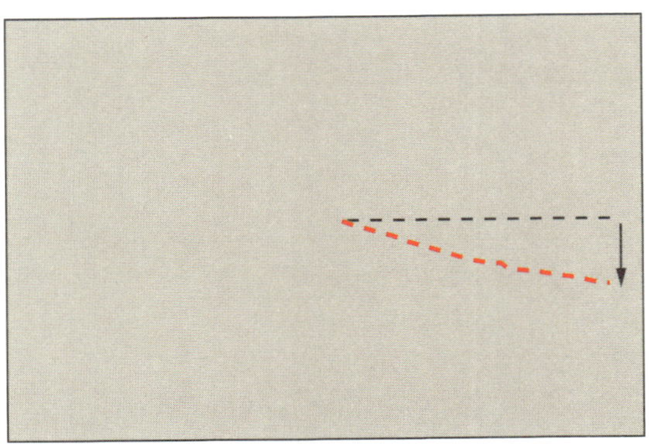

figure 3–1d *Interpretation*

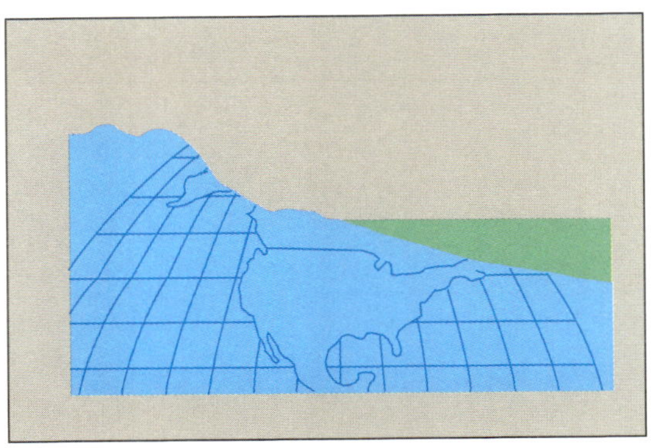

figure 3–1f *Embellishment*

Figure 3–2 *A second example of a complex illustration separated into its component parts. Note that this illustration does not contain interpretation.*

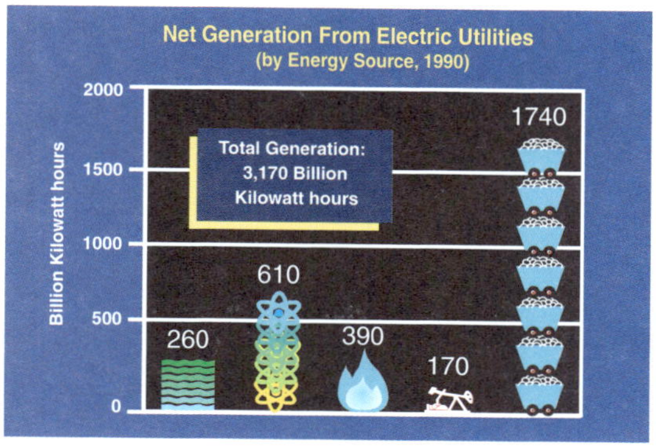


figure 3–2a *Composite Illustration*

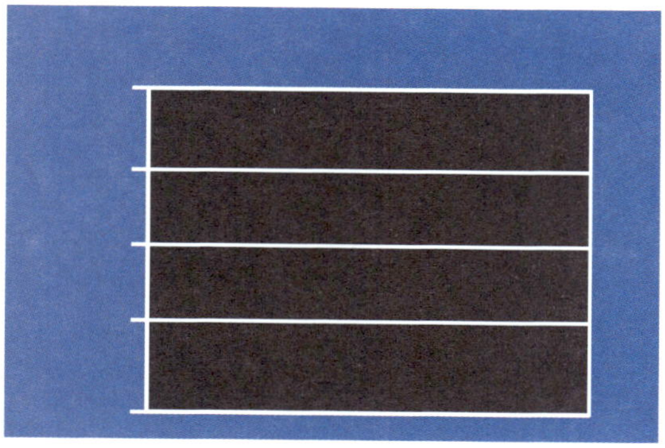

figure 3–2b *Framework*

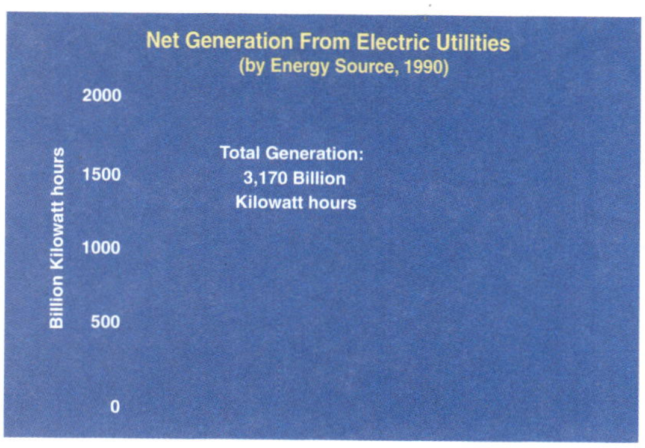


figure 3–2c *Annotation*

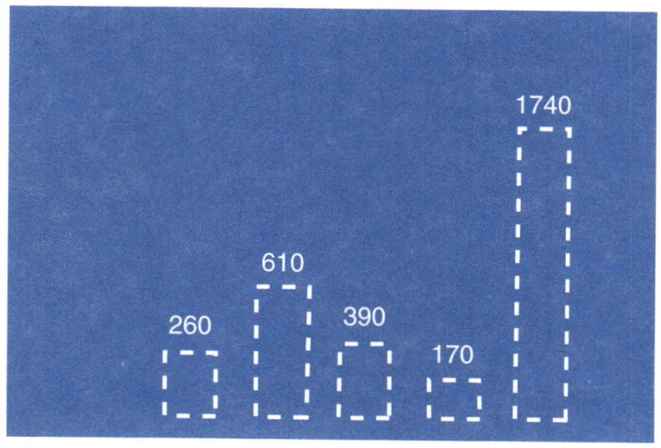


figure 3–2d *Data (height of bars)*

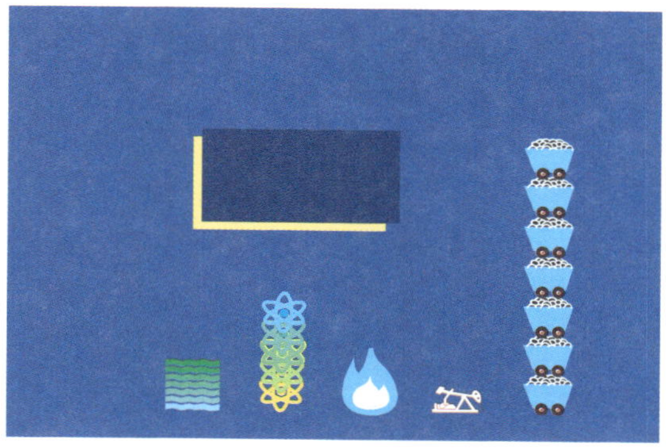

figure 3–2e *Embellishment*

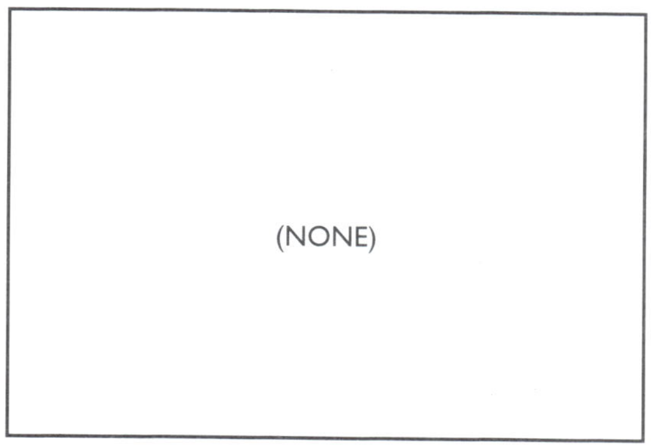


figure 3–2f *Interpretation*

The framework is the grid of the chart, and the data are the heights and numeric labels of the bars. There is no interpretation (remember, not all illustrations have all parts), and annotation includes the labels on the axes, the title, and the summation of total generated electricity. The replacement of traditional bars by icons of the fuel types constitutes embellishment.

Each of these examples is relatively well designed for their individual purposes. Figure 3–1 is more flamboyantly embellished (to "sell" a political program), and Figure 3–2 has a less embellished, more utilitarian appearance.

Each also has elements that could be discarded without reducing the amount of information conveyed. Specifically, the map on Figure 3–1 adds no relevant information, and the heights of the bars and numerical annotation of kilowatt hours on Figure 3–2 are redundant (although the numbers add precision).

## *The Layers of an Illustration*

The definition of parts of an illustration is similar to the drafting technique of constructing a drawing from a series of overlays or layers. In the days of mechanical or board drafting, it was a common practice to create a complex illustration by drawing sets of information onto a series of clear mylar sheets, each registered so that the whole series could be stacked to form the composite drawing. Layering allowed easy editing, separation of colors for printing, and the option of selectively adding or subtracting layers when the illustration was used for different purposes.

This same concept has been extended to computer graphics and drafting programs. Many computer graphics programs allow segregation of graphical elements onto separate electronic layers, which can then be turned either on or off depending on the immediate use of the drawing.

Taking this concept into a consideration of the parts of an illustration, it is easy to visualize framework, data, interpretation, annotation, and embellishment as separate layers, each building on the others to create the complete illustration (Figure 3–3). Also, as we illustrate below, more than one part may be combined on a single layer.

figure 3–3 *The "parts" of an illustration depicted as a series of "layers" that stack into a composite illustration*

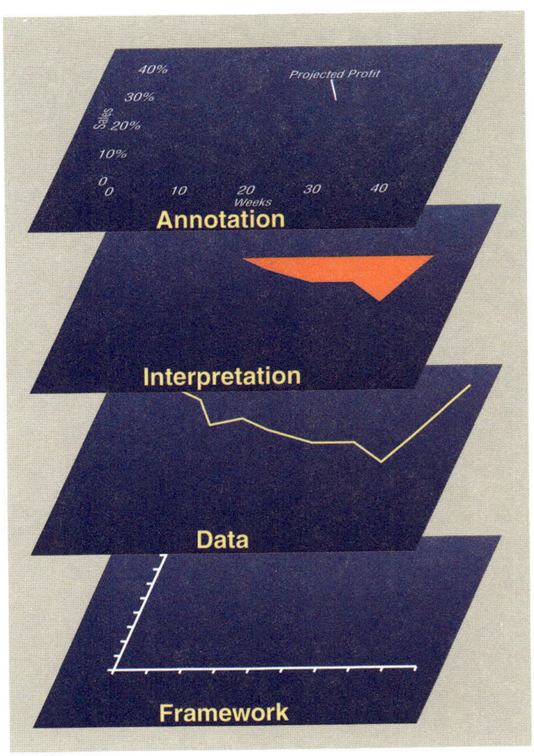

## *Why Does Any of This Matter?*

Theory is nice, but what use does it have for professional presentations? To answer that, let's look at an example of an illustration that appeared in a professional publication (Figure 3–4). As is, this illustration would make a very poor slide. It is cluttered and confusing. Simplicity, clarity, and legibility are nowhere to be found. However, let's apply the lessons of this chapter to see what can be done.

Figure 3–4a is the original drawing, with a great deal of overlapping and confusing detail. In Figures 3–4b–e, this illustration has been broken into its component parts. In Figure 3–5, these parts are recombined into layers. The primary or base layer consists of framework (the axes of the graph), data (the plotted data values), and annotation (the labels on the axes and the annotation of data points). In fact, this layer creates a simple, clear, and legible illustration.

figure 3–4 *This original, highly complicated illustration has significant problems and is wholly inadequate for a slide. Here, we have separated it into its components in preparation for better presentation of the material.*

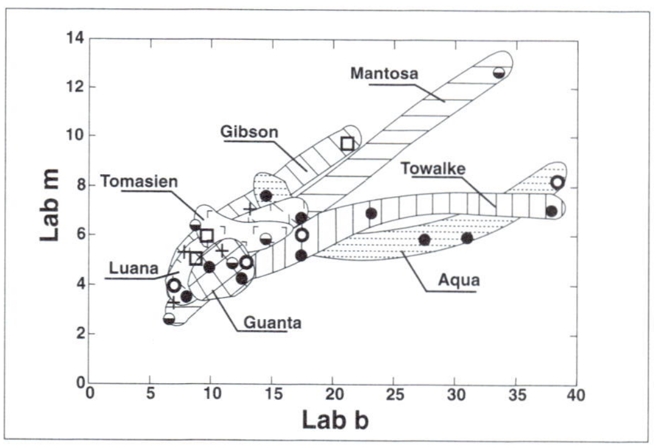


figure 3–4a *Composite Illustration*

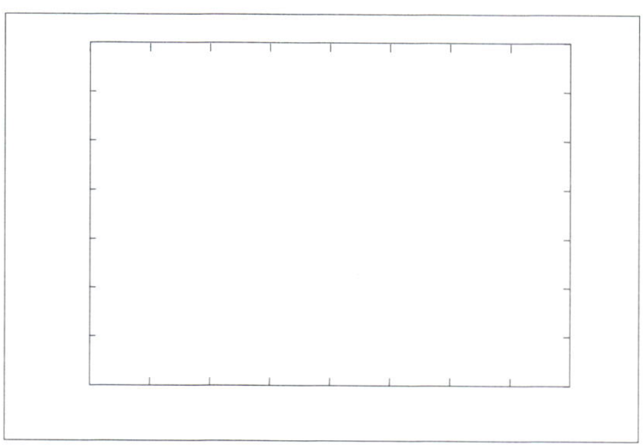

figure 3–4b *Framework*

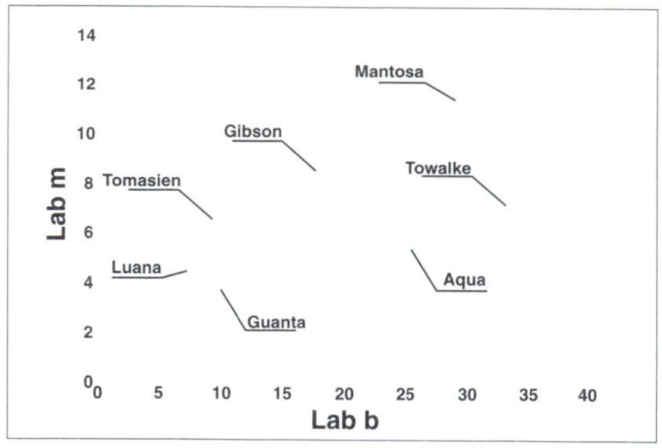


figure 3–4c *Annotation*

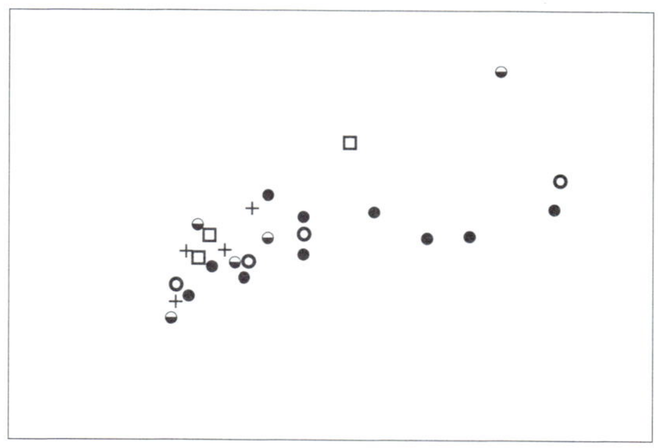

figure 3–4d *Data*

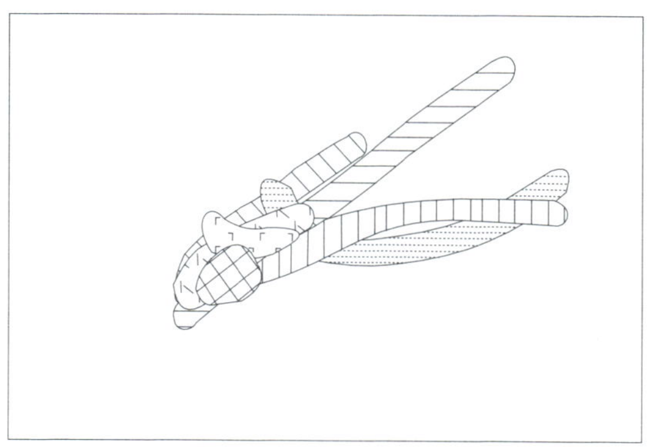

figure 3–4e *Interpretation*

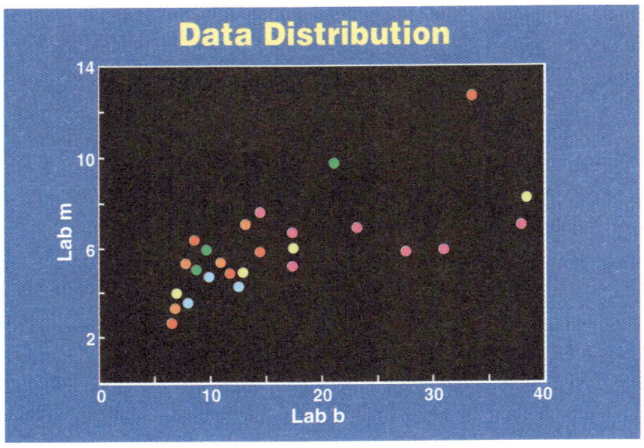


figure 3–5a *Base*

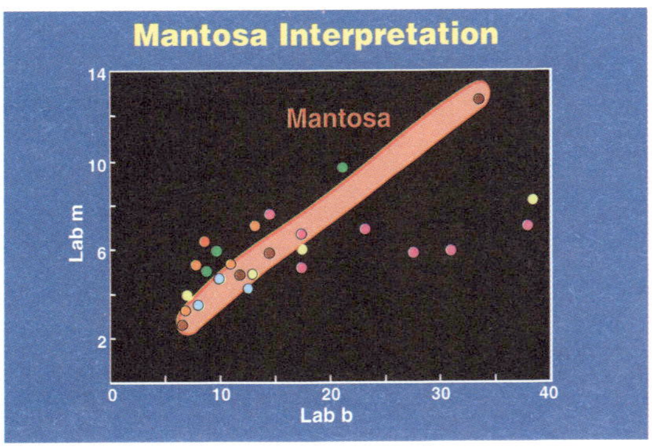


figure 3–5b *Base with Mantosa interpretation*

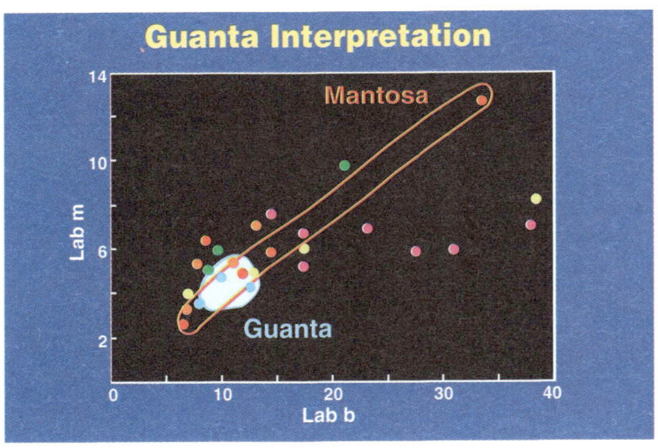


figure 3–5c *Base with Guanta interpretation*

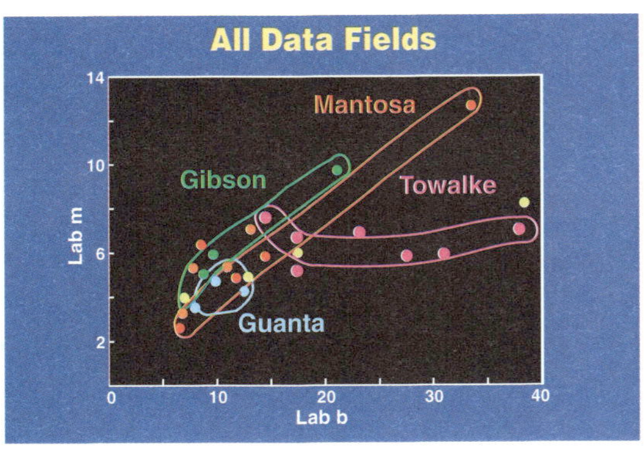


figure 3–5d *Base with four of the data fields identified*

Clearly a problem arises with the addition of all of the interpretation (the circled groupings of data points) and the stippling. If interpretation and embellishment must be shown, a reasonable solution would be to create a separate layer for each data type (e.g., Guanta), add a unique caption or title (part of the annotation), and depict these one at a time in a series, reusing the same base for each (Figure 3–5b,c, and d).

This example shows the value of analyzing illustrations for their component parts and how those parts can be layered for presentation. This approach is particularly important when illustrations are visually complex or crowded and the format (e.g., 35-mm slides) imposes constraints on the illustration.

A few helpful guidelines for slide design arise from this approach:

- Discard unnecessary and distracting embellishment
- Reduce annotation to a minimum
- Segregate layers onto more than one slide. Too often presenters use a complex illustration instead of two or three simple ones.
- Lead viewers through complex illustrations by successively adding layers in a sequence of slides, or by adding information through animation in electronic presentations

As you read the next chapter on slide design, look back at Figures 3–4 and 3–5 to see how easily a complex figure can be made simple, clear, and colorful—a critical process for turning illustrations into slides.

figure 3–5 *This sequence of illustrations demonstrates how the detailed diagram from Figure 3-4 can be improved by simplifying, rearranging the components, and breaking a single illustration into a series. The illustrations have also been improved by the use of color to define data points (rather than simply for embellishment).*

# 4

# *Slide Design—The Basics*

All sciences make use of graphics to communicate data, interpretations, and ideas. Our journals are full of these illustrations, and we illustrate oral presentations with slides. Unfortunately slides—and the time constraints normally imposed on the presenters—severely limit the amount of graphical information that can be shown effectively. This leads to the great importance of proper design for excellent slides.

## Simple, Bold, and Colorful

The goal of a speaker is to have a set of slides that enhance a talk. They should blend into a talk, logically following the sequence of ideas and supporting and illustrating the speaker's statements. In a good talk, the audience absorbs the oral and visual information equally. Conversely, just as a successful set of slides will enhance a speaker's attempt to convey information, poor slides will hinder the effort.

We have all seen good slides, and we have seen bad slides. What makes them either good or bad? Bad slides are poorly laid out, hard (or impossible) to read, and cluttered with unnecessary information. They flash by, and the audience has no time to figure them out. As that happens, the audience loses interest or becomes frustrated; chairs shuffle and geese honk.

**Purposes of Slides**

- Clearly impart information
- Enhance the verbal presentation
- Keep the speaker on track

Designing a slide is very different from designing a graphic for other purposes. Too often, slides are created or photographed from illustrations prepared for print, without the essential redesign for the slide format. This is rarely satisfactory. Slides have a fixed size format and must be legible when projected. Furthermore, slides flash past a viewer's eyes at a pace—often a very rapid pace—controlled by the speaker. The viewer is not in control of the amount of time available to absorb the information. All of this places serious constraints on the design and content of a good slide.

Let's look at a well-designed slide (Figure 4–1). It is simple; it is bold; and it makes good use of color. It clearly displays its information. It will project well, and the audience will grasp its purpose at a glance.

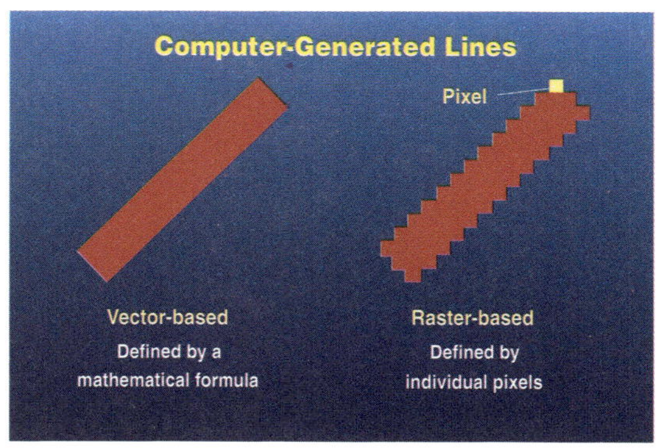


figure 4–1 *An example of a well-designed slide*

These three concepts—simplicity, boldness, and use of color—are guidelines for producing professional-quality slides.

| | |
|---|---|
| Simplicity | A good slide shows only one thing—the information the speaker is presenting at the time it is projected. It supports one major point. It is uncluttered, and anything that does not support the point is discarded. |
| Boldness | All graphics and text are clearly visible and legible to the audience throughout the room. |
| Color | Colors, including the background color, highlight, clarify, and contrast features. All together, they produce a visually pleasing image. |

## What is on a Slide?

Slides may be created to communicate any conceivable type of information. Although this book focuses on technical and scientific presentations, the principles apply all the way from rocket science to children's entertainment. Regardless of the specific content, all slides comprise some combination of background, text (annotation), graphical illustrations, and/or photographs. And, all slides must be properly laid out.

figure 4–2 *Font sizes and line thicknesses (The font and line sizes illustrated here were typed on a full-size template and are scaled proportionate to the size of the illustration.)*

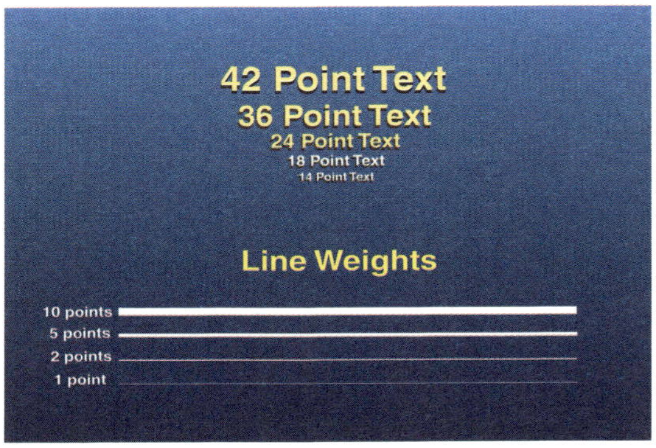


figure 4–3 *Examples of contrasting colors and text on colored backgrounds. Notice the need for strongly reversed text colors.*

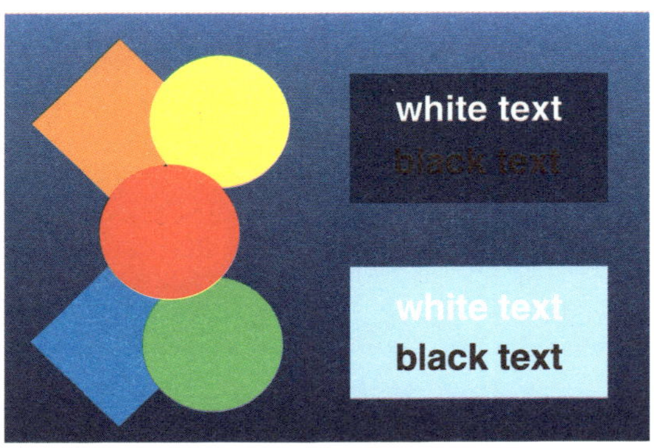


The following is a helpful set of guidelines.

Background

- Use a colored background, not white.
- Blue has become the most popular background color; black and green are also common.
- Background colors can be gradational; grade colors from dark at the bottom to light at the top.
- Background embellishments, such as color bars running down one side, watermarks of logos, etc., may reduce the area and/or readability of the image.

Text

- 18-point (Figure 4–2) text is the smallest that will project (rarely, non-essential text such as an acknowledgment is included in a smaller font size).
- 36-point text is the largest you will ever need.
- The use of upper and lower case letters (rather than all caps) aids readability.
- Use sans serif fonts. (Some serif fonts are difficult to read if the letters vary widely in thickness.)
- Avoid flowery fonts and script.
- Use appropriate text colors (Figure 4–3) to stand out from the background (e.g., yellow on blue, white on black). Avoid red text because it is difficult to read.

Graphical Illustrations

- Draw illustrations in the simplest possible form while retaining the necessary information. Discard any information that is extraneous to the point.
- Line weights of one point or two points are appropriate for most framework, outlines, and data (Figure 4–2).
- Use colors to separate and clarify graphical elements from each other and from the text (Figure 4–3).

Photographs

- Photos may fill the entire slide or be embedded into a background, with annotation or graphics as appropriate.
- Embedded photos should occupy no less than one-quarter of a slide (larger if there is much fine detail).

Layout

- If at all possible, lay out slides in a horizontal (landscape) mode for assurance that they won't project over the top of the screen or be cut off by audience shadows.
- Choose a template with appropriate proportions (height to width) for a slide. The best custom page for most software is 7 1/3 x 11 inches for a slide and 7 1/2 x 10 inches for an electronic presentation. The dimensions 8 1/2 x 11 inches will not work.
- Fill the available space (see comment above about embellishing the background).
- Create templates for any series of similar slides.
- Leave a margin of one-quarter inch on all sides for the frame of 35-mm slides.

## *Using Templates for Consistency*

Templates are style guides for the design of slides. They specify location of titles, graphics, logos, and other objects. They also specify backgrounds, font sizes, bullet styles, and colors. The use of templates speeds up the process of designing slides. It also creates a consistent appearance for all of the slides in a talk, which increases the effectiveness and professionalism of the presentation.

Figure 4–4 presents examples of templates. Notice that background elements can limit the space available for the information content of the slide. The examples show that choices for the placement of a logo and other elements such as color bars require attention to this space issue.

Software programs often have several pre-designed templates from which you can choose, or you can design and save a template of your own. Many major corporations specify templates for all presentations made by their employees.

figure 4–4 *Examples of templates for slide design. Red boxes on these slides segregate areas reserved for titles, text, logos, and graphical elements. Text formats can be defined for each area. The boxes do not appear on the finished slides.*

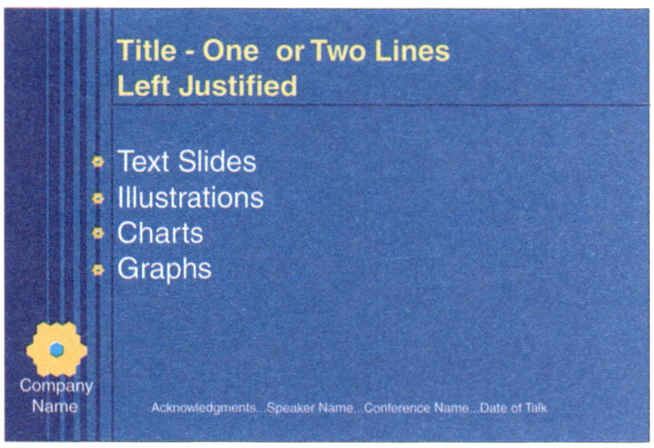


figure 4–4a

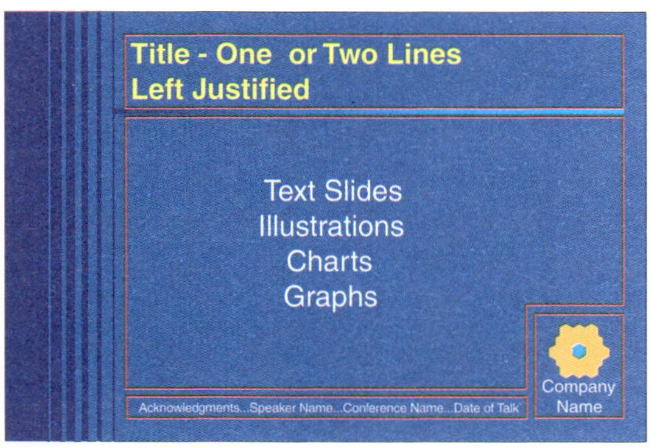


figure 4–4b

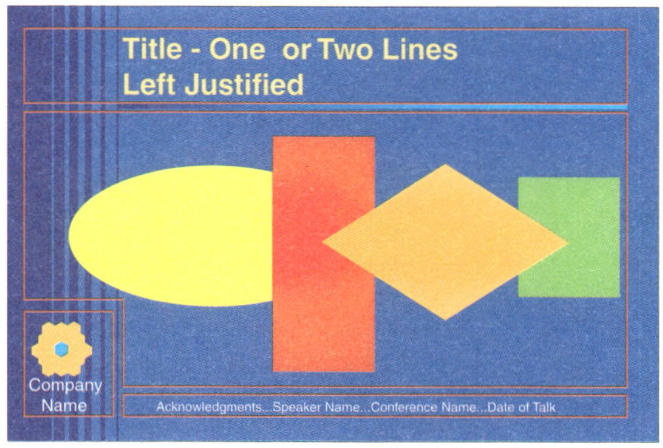


figure 4–4c

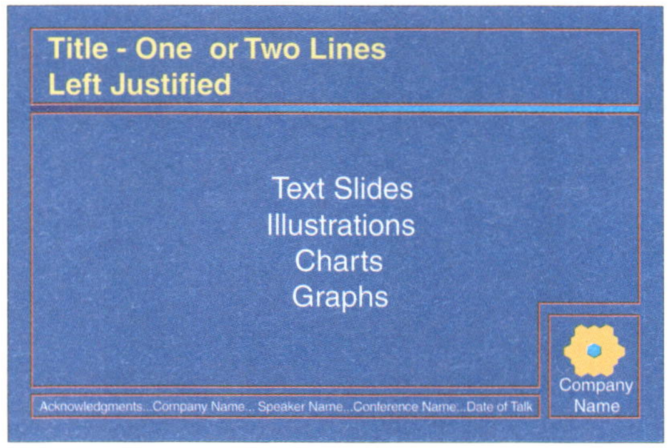


figure 4–4d

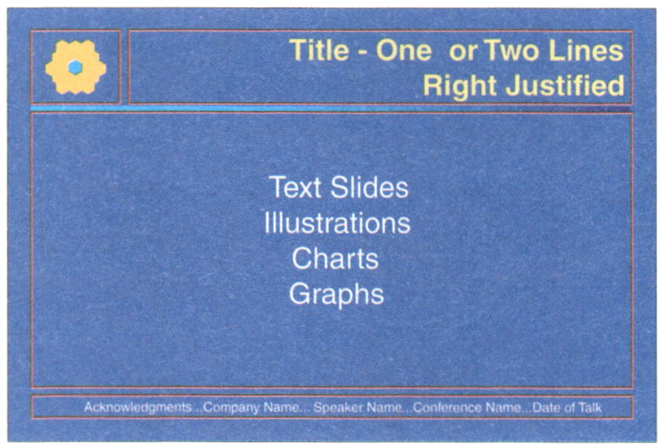


figure 4–4e

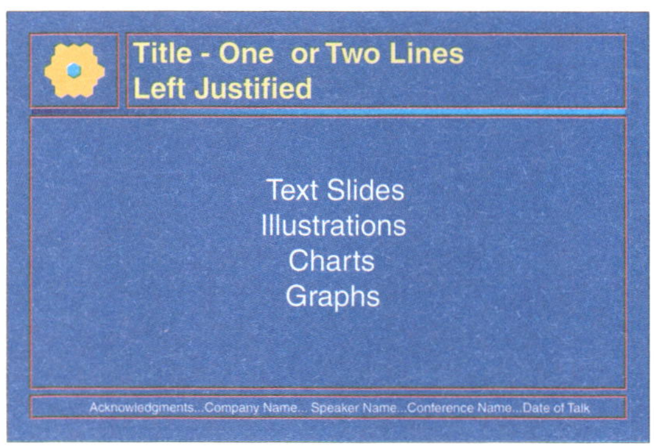


figure 4–4f

## *Quick Checks for Legibility*

How can you tell if a slide will be legible? Here is a "quick check technique." Make a letter-sized paper print, pin it on a wall, step back five or six feet, and read it. If everything is readily legible at that distance, it will project legibly. Similarly, if you are designing slides on a 17-inch monitor, text and features that are readily visible from a distance of eight to ten feet will project well.

Another quick check is to look at the monitor from a normal reading distance, cast your eyes slightly out of focus, and look to see if the major points still stand out. If not, they may be illegible when projected.

## *Simplification for Legibility*

As we have already emphasized, 35-mm slides provide a very restrictive format for depicting graphical information. To achieve simplicity and boldness, we focus each slide on a single major point, we use the fewest words possible, we leave off non-essential graphical information, and we redraft complex illustrations into their simplest possible form. Remember, a slide stays on the screen for an average of only 30 seconds—the audience must comprehend the information very quickly.

This need for simplification often comes about when designing slides from illustrations that have been prepared for hardcopy publication. Printed illustrations can include highly detailed information, small fonts, and subtle features. Slides cannot. To redraft legible slide copy from a printed figure, two techniques are applicable. The first is to reduce the amount of information shown, and the second is to use a series of slides to depict the full amount of information. We have already seen examples of these techniques in Chapter 2. Let's look at another.

Figure 4–5a, reproduced from a professional journal, provides information about the duration of the Southwest Monsoon in various regions of India. As is, it would make a very unsatisfactory, "busy," and illegible slide. Here are two appropriate solutions for depicting this information on slides.

The first solution is to present the onset and withdrawal dates of monsoon conditions separately (Figures 4–5b and 4–5c). These could be projected on side-by-side screens or sequentially. The second solution is to present the onset and withdrawal information in the form of duration zones (Figure 4–5d), yielding a highly simplified slide. The choice between these solutions would be based entirely on the purposes of the speaker.

Notice the other ways in which these slides have been simplified. There is no north arrow (north at the top of a map is conventional), not all contours are annotated, we left off the scale (the Indian subcontinent would provide sufficient scale for many purposes), and we left off latitude and longitude. Of course, the speaker could keep any of this omitted information on the slides if it were important.

### A Caution About Corporate Standards

Many corporations carefully protect the use and identity of their logos. They publish usage standards in style guides and zealously guard against misuse. In fact, "logo police" is a common appellation for those who enforce the standards.

These guidelines often precisely specify colors, placement on slides or viewgraphs, and distance from other graphical elements for the use of a logo.

These specifications may preclude the use of certain background colors or other features of a template. Knowing the standards in advance will eliminate accidental misuse of a logo or the need to redesign slides.

figure 4–5 *An example of simplification. The original diagram (a) is too complicated to depict in its entirety on a single slide. Instead, show the set of dates for onset on one slide (b) and withdrawal on another (c) or, replace those two slides with a single slide that presents the seasonal duration of the monsoon (d).*

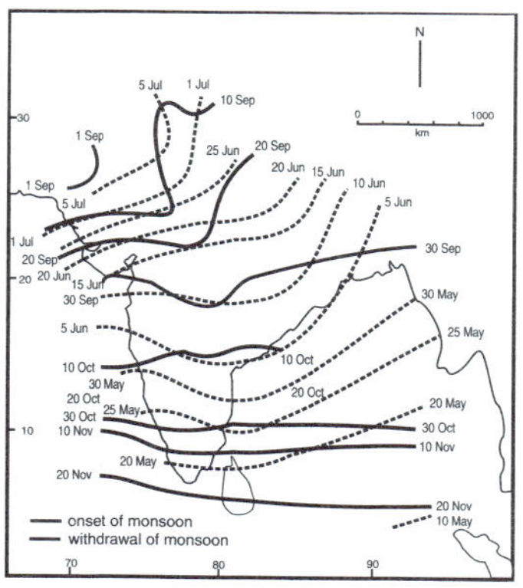


figure 4–5a

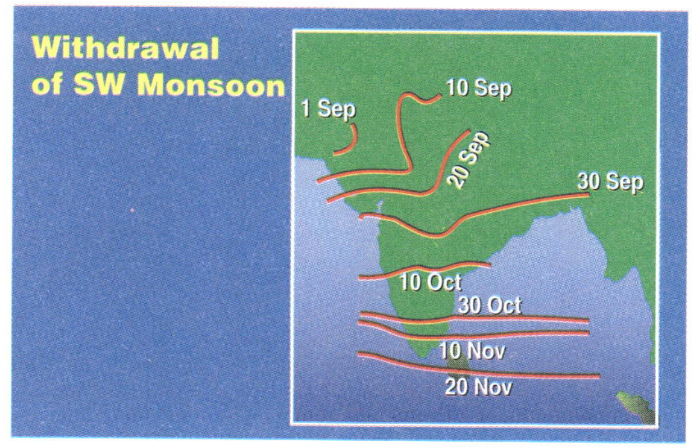


figure 4–5c

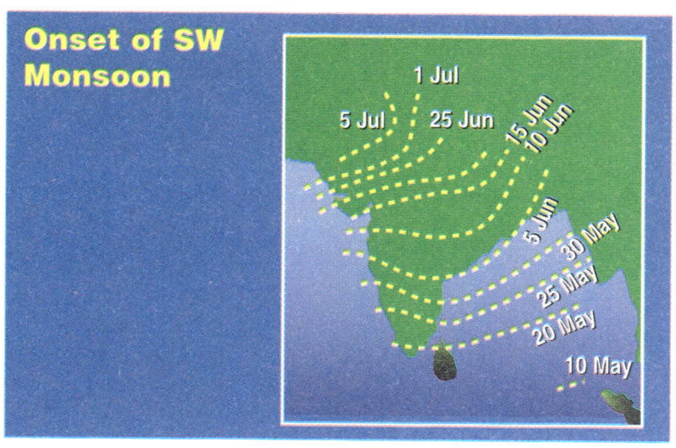


figure 4–5b

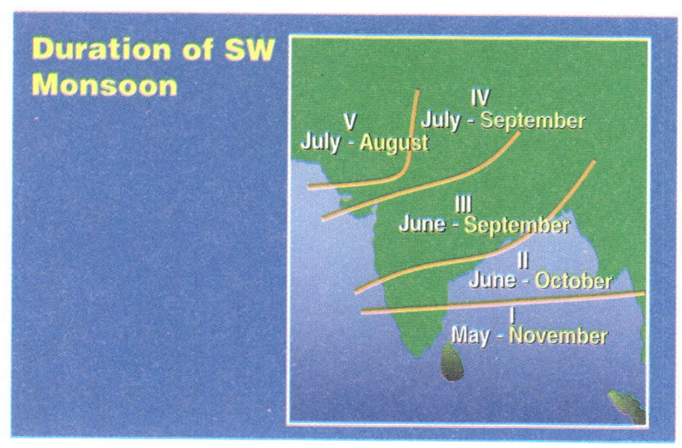


figure 4–5d

# 5

# *Tricks of the Trade*

Experienced slide designers have developed practices and preferences to ensure that their slides are simple, bold, and colorful. In this section, you will find several of these tips. And, we have an expert goose on hand to judge whether an example is :

unacceptable –

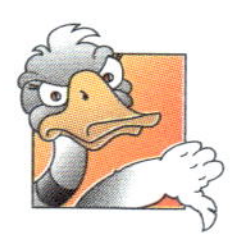

or

excellent –

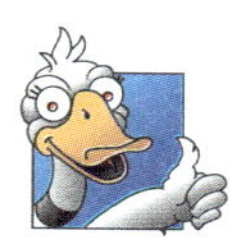

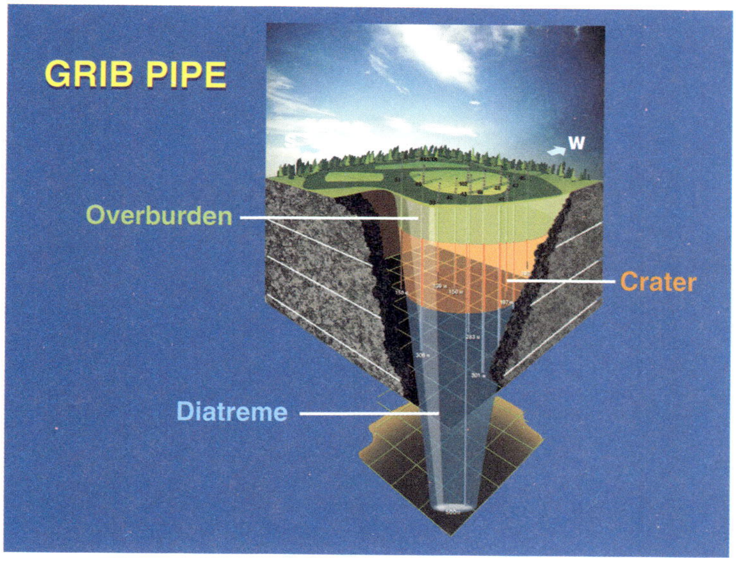


## *Simple, Bold, and Colorful*

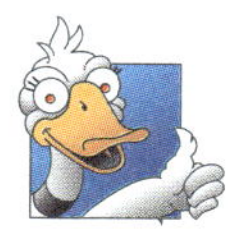

figure 5–1 *This slide demonstrates the themes of simplicity, boldness, and color. The illustration is not overly complicated, the text is easily legible, and the viewer is immediately engaged in understanding the geologic structure, not the structure of a complex slide.*

## *Build to a Conclusion*

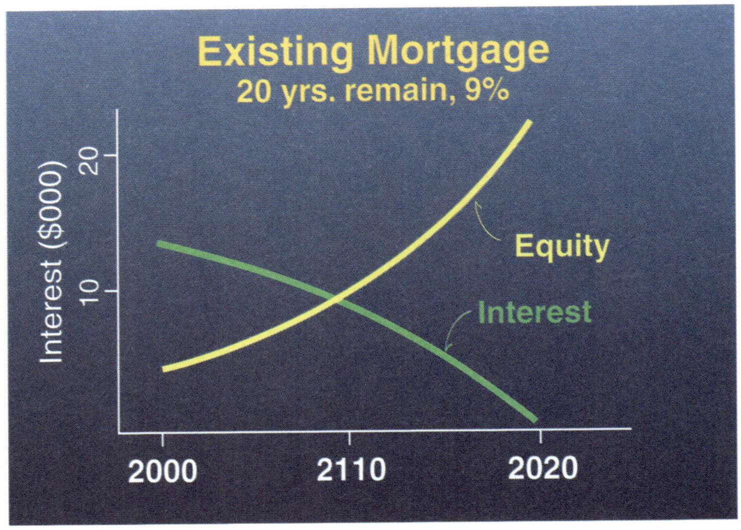


figure 5–2a

This series of four slides builds to a conclusion—that a homeowner can save a lot of money by refinancing a mortgage to a lower interest rate and shorter term. As the slides build, notice how previously presented information is screened, and new information is prominent. Refer to Figure 3–5 for another example of using a series of slides to build to a conclusion.

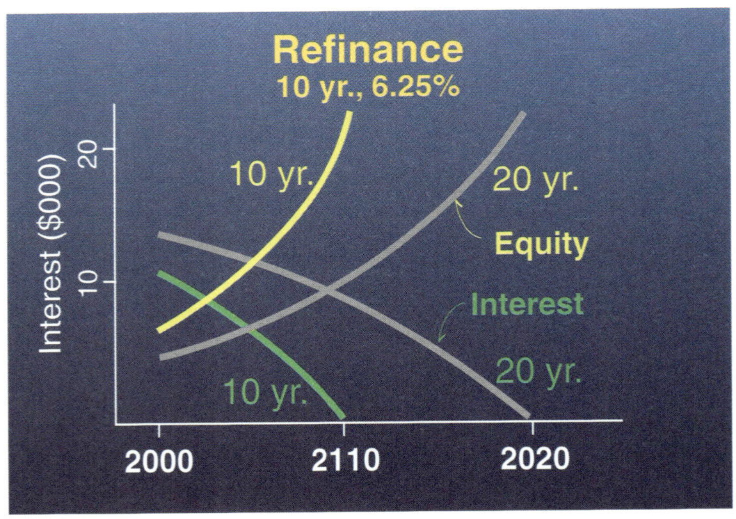


figure 5–2b

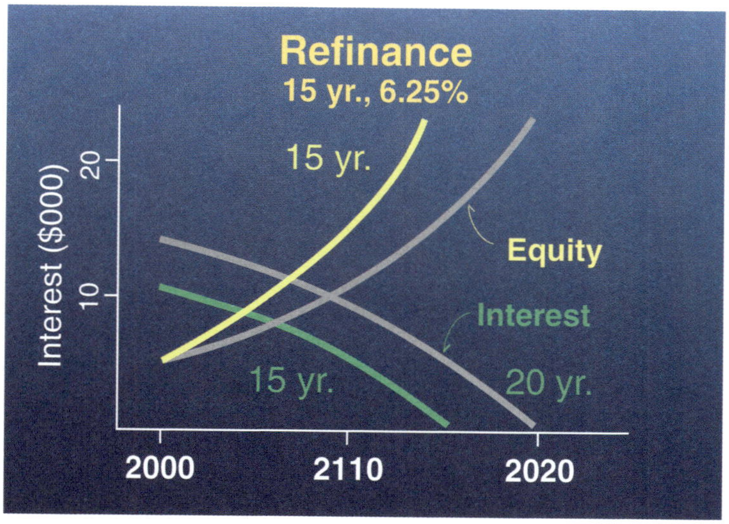


figure 5–2c

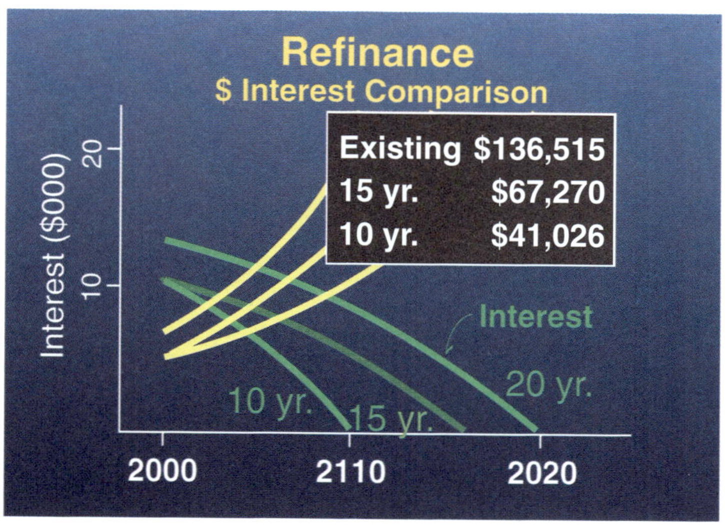


figure 5–2d

## *Reduce Framework to a Minimum*

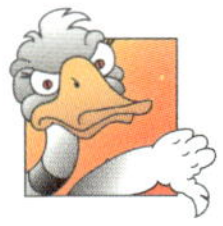

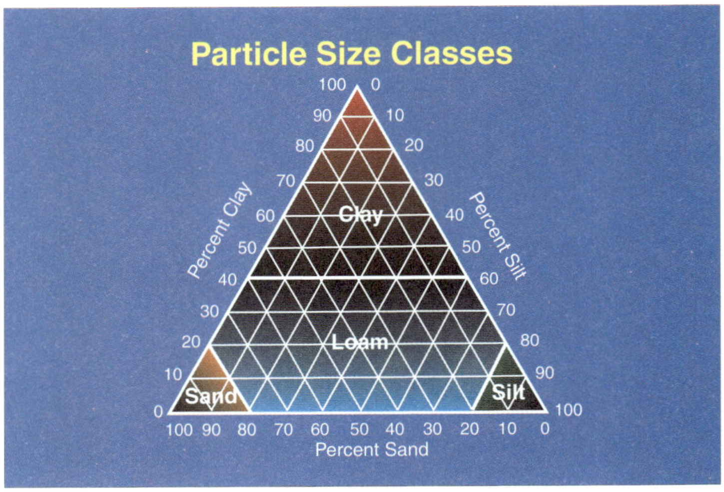


figure 5–3a

The excessive detail in the framework of this ternary diagram (Figure 5–3a) implies that the audience will want (and have time) to work out exact percentages of constituents for the various classifications. They won't. A much simpler framework is shown in Figure 5–3b. (Notice that the annotation has become more readable.)

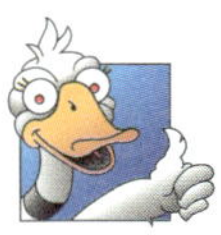

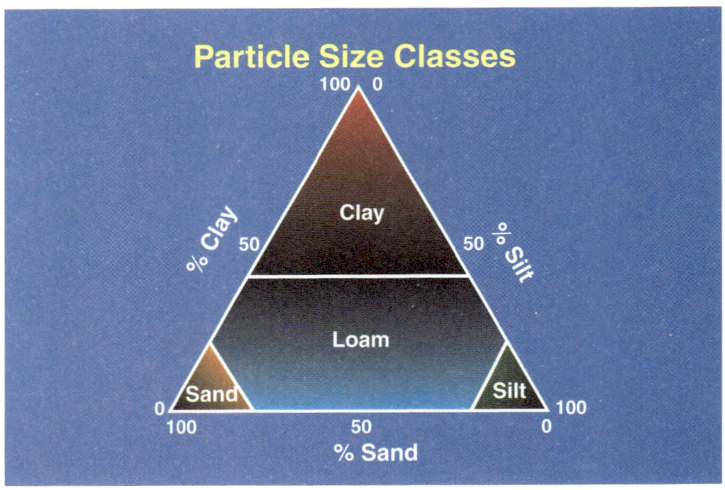


figure 5–3b

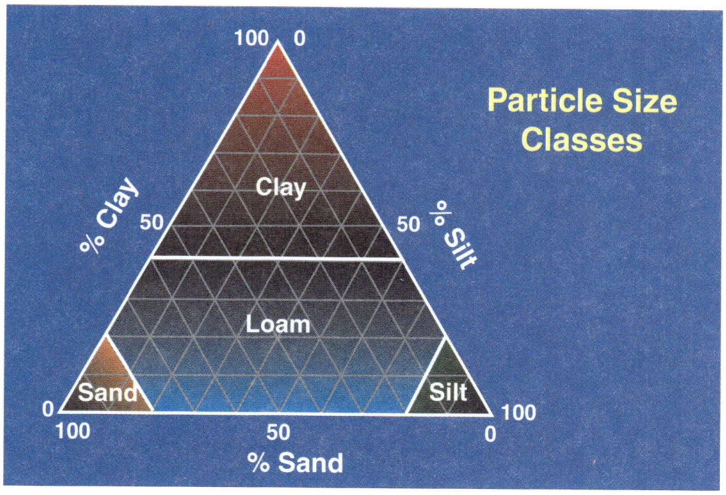


figure 5–4

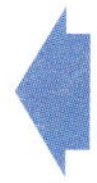

*Subdue Framework into Background*

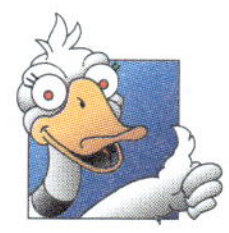

Reduce the visual dominance of framework by putting it in gray or another subtle, light color. This technique preserves the entire framework of Figure 5–3a, but the grid interferes less with the annotation.

## *Use Embellishment for Enhancement*

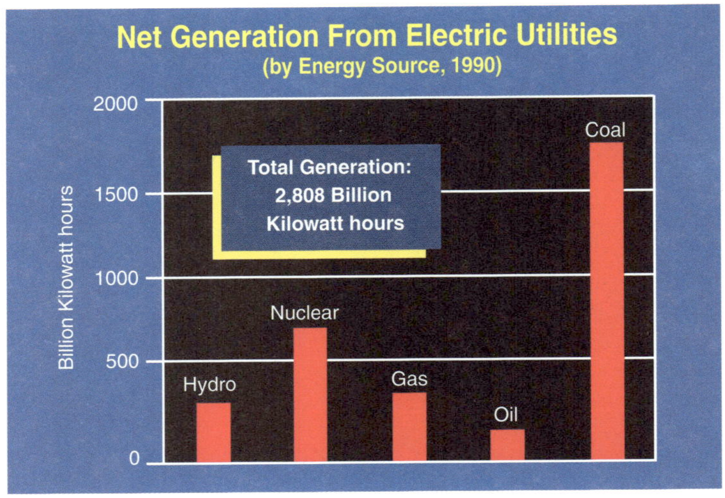


figure 5–5a

The slide of the bar chart in Figure 5–5a is nicely designed. Additional embellishment in Figure 5–5b enhances the graph, adds visual impact and appeal to the audience, and allows a reduction in the amount of annotation.

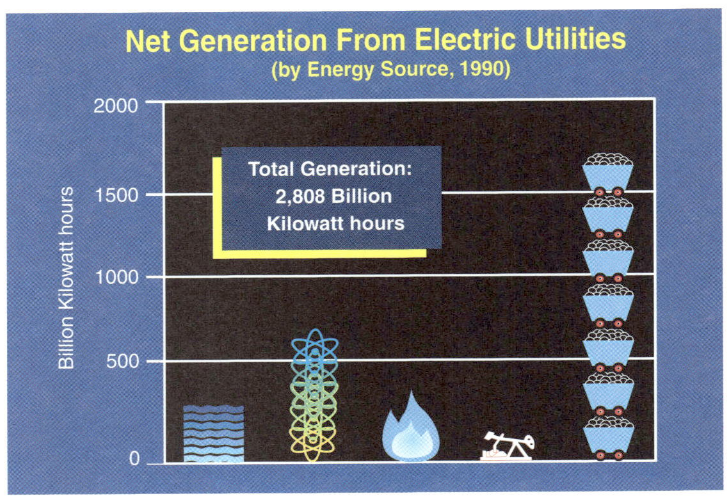


figure 5–5b

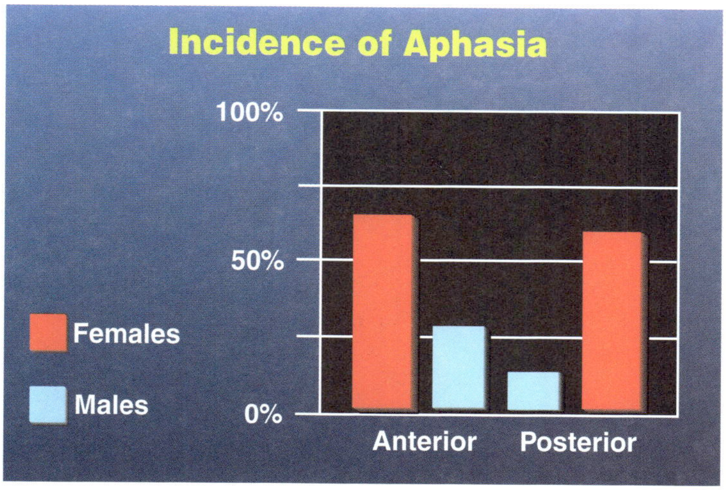


figure 5–6a

Note how uninteresting Figure 5–6a is. The superposition of this bar chart over a drawing in Figure 5–6b allows the audience to immediately see the relationship between the data and the areas of the brain, yet the slide remains legible and uncomplicated.

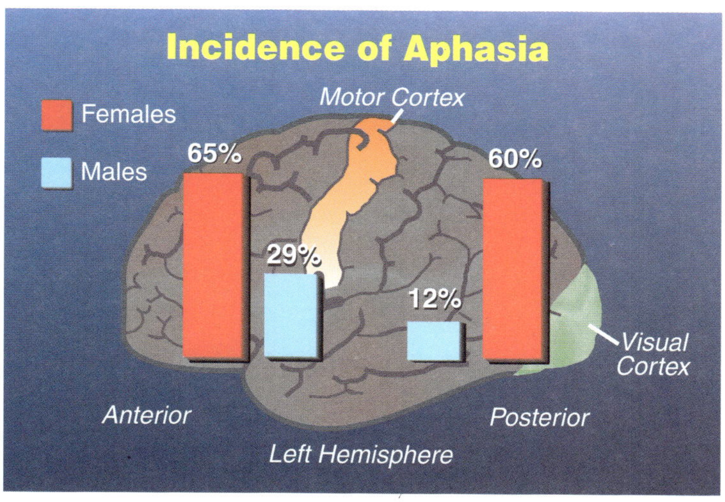


figure 5–6b

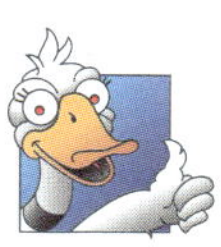

# *Eliminate Unnecessary and Distracting Embellishment*

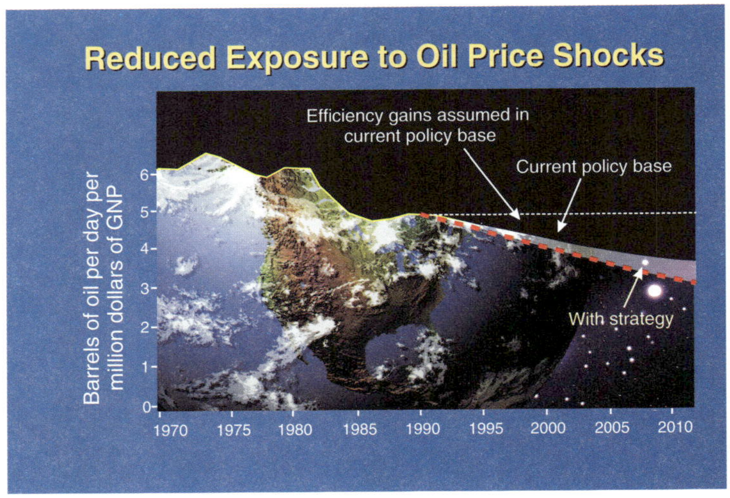


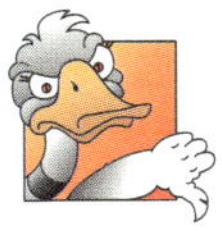

figure 5–7a

The partial view of the globe in Figure 5–7a overwhelms all other elements of this line graph without imparting any relevant information. The simple embellishment of Figure 5–7b interferes less with the data and interpretation. Unnecessary labeling has also been removed.

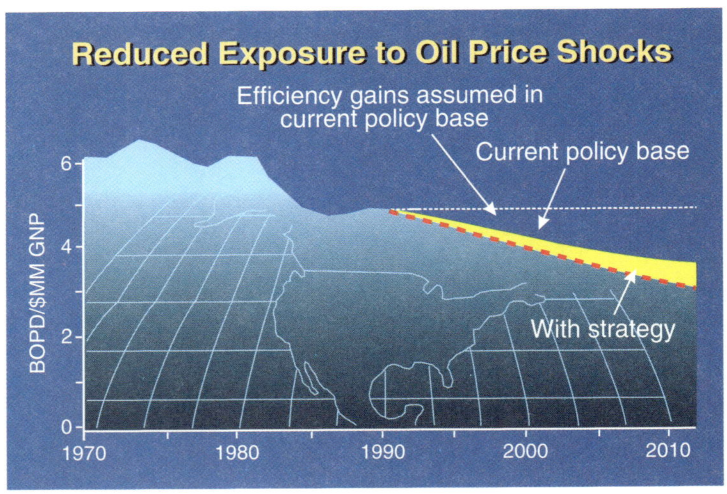


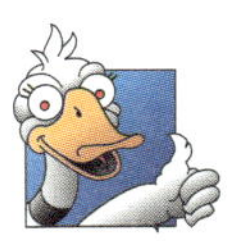

figure 5–7b

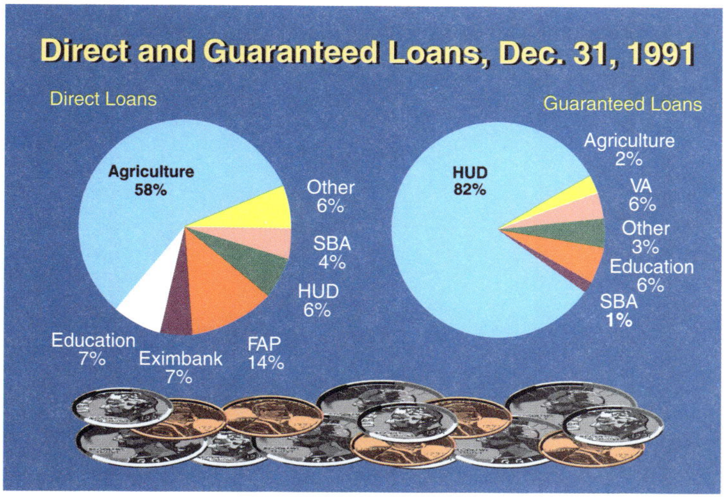


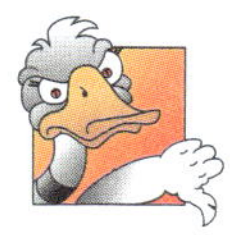

figure 5–8a

What good are the coins in Figure 5–8a? They distract the viewer and take up space, causing the artist to reduce the size of the pie diagrams. A somewhat better use of the same embellishment is to create pie charts out of coins. Notice the increased amount of space on the slide available for annotation.

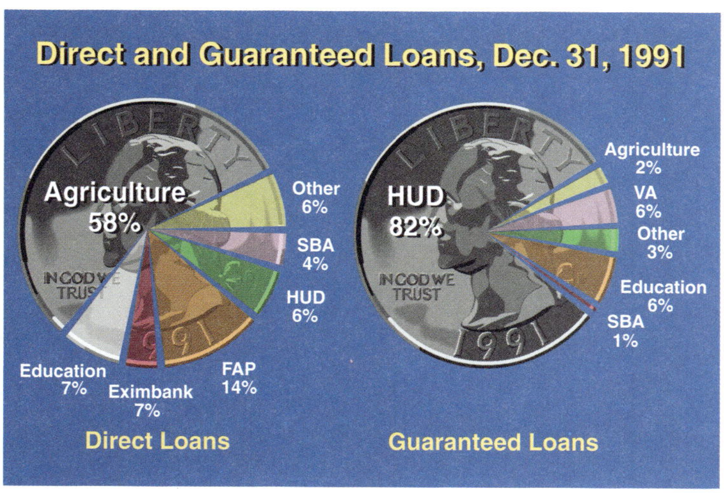


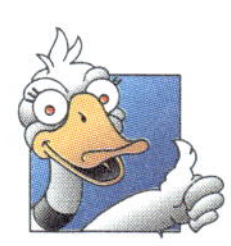

figure 5–8b

## *Make Efficient Use of Available Space*

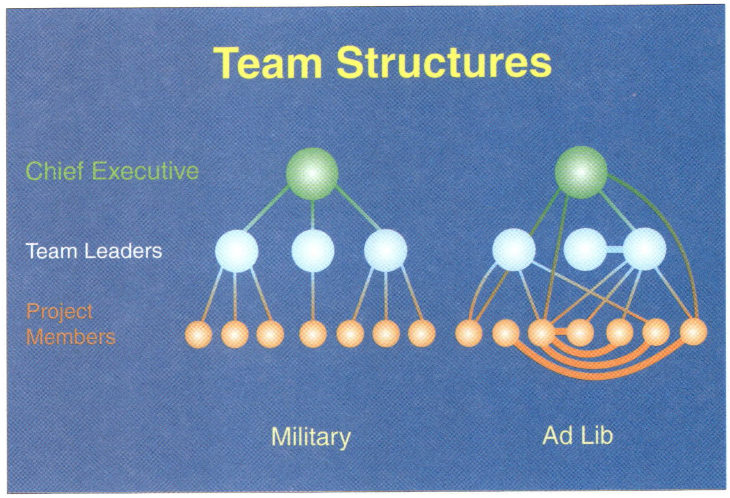


figure5–9a

Annotation on the left side of Figure 5–10a wastes space and causes the artist to reduce the size of the figures. Bringing this annotation to the center of the slide is a big improvement. For another example of the efficient use of space, notice the larger size of the ternary diagram in Figure 5–4 in comparison to Figures 5–3a and 5–3b.

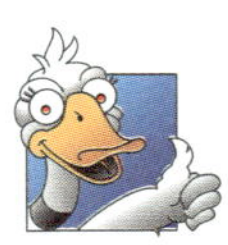

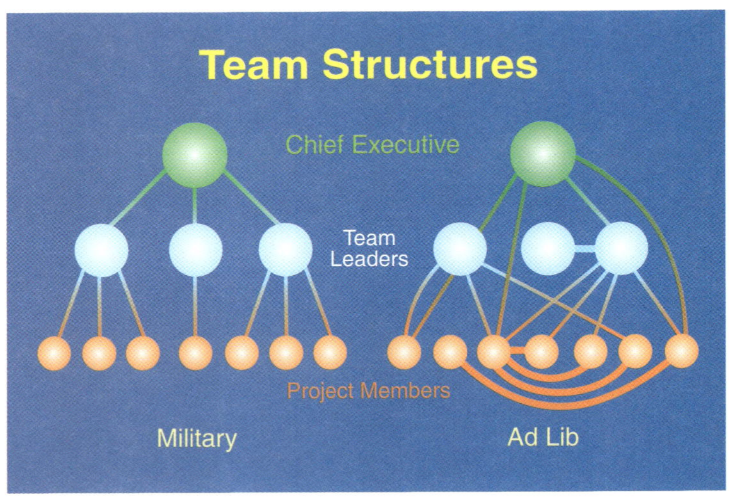


figure 5–9b

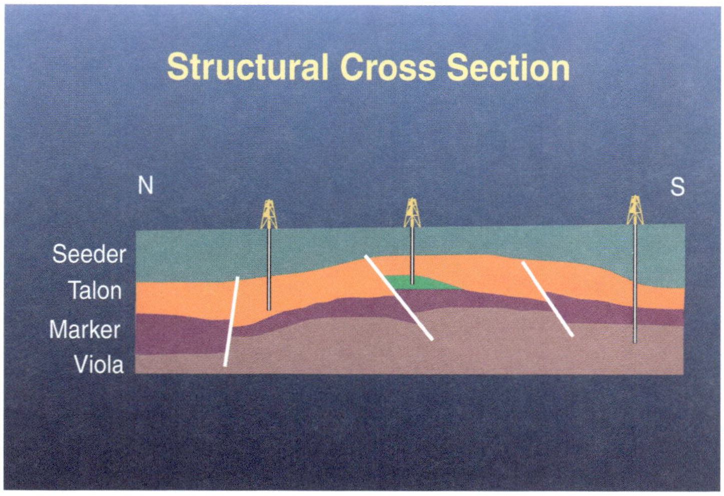


figure 5–10a

Direct labeling of intervals on this geologic cross-section allows the efficient use of space and enlargement of the cross section.

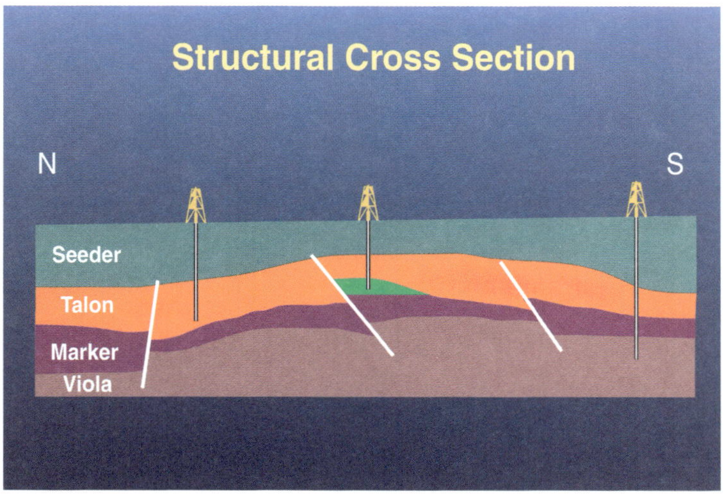


figure 5–10b

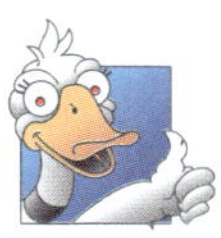

## *Word Slides*

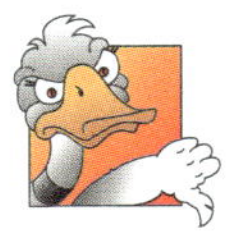

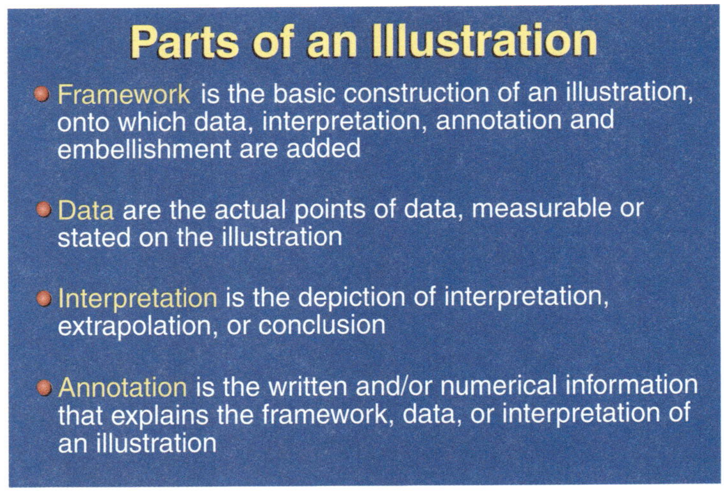


figure 5–11a

Fewer words are better, as these examples illustrate. Figure 5–11a is much too wordy. Figure 5–11b reduces the sentences to statements. For a two-screen presentation, Figures 5–11c and d work together to show the list of the parts and the complete definition of one of them. Figures 5–11b–d follow the rule of thumb of having no more than six lines of text.

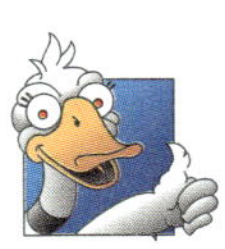

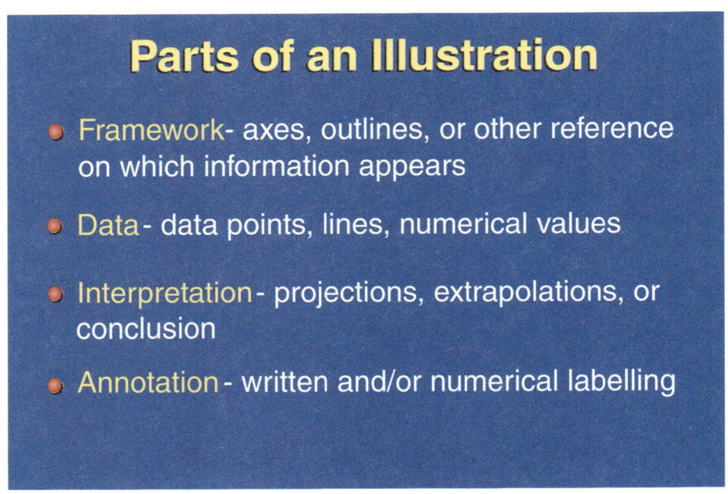


figure 5–11b

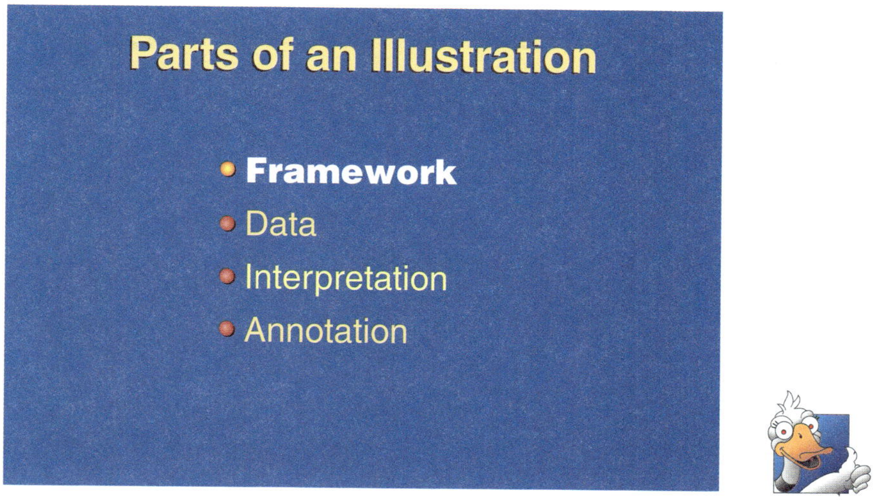


figure 5–11c

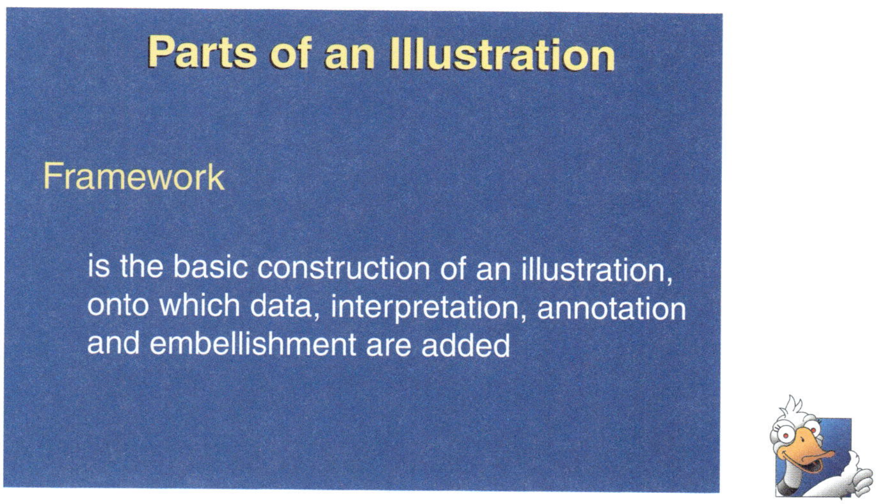


figure 5–11d

## *Working With Tables*

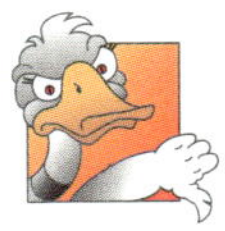

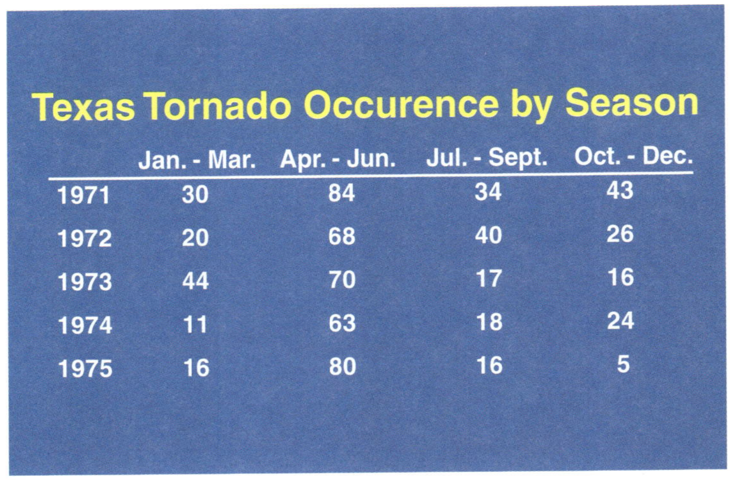


figure 5–12a

Tables present special challenges for slide design. Even a fairly simple table such as Figure 5–12a is overcrowded with information, and the information is difficult to grasp in a quick glance. Two solutions are illustrated here. In Figure 5–12b, a colored box highlights specific information to draw the audience's eyes. Better yet, Figure 5–12c presents the same information graphically, which is a clearly superior means of communicating the information.

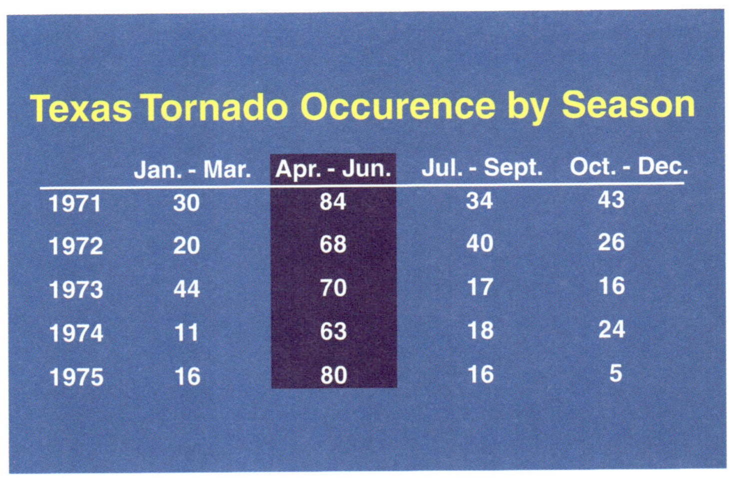


figure 5–12b

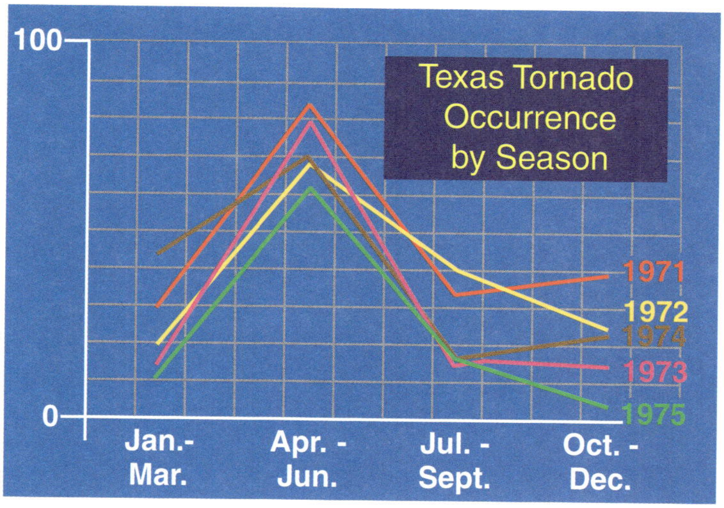
100
0
Texas Tornado
Occurrence
by Season
1971
1972
1974
1973
1975
Jan.-
Mar.
Apr. -
Jun.
Jul. -
Sept.
Oct. -
Dec.

figure 5–12c

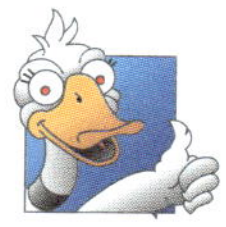

## *Use Side-by-Side Screens Effectively*

Left Screen

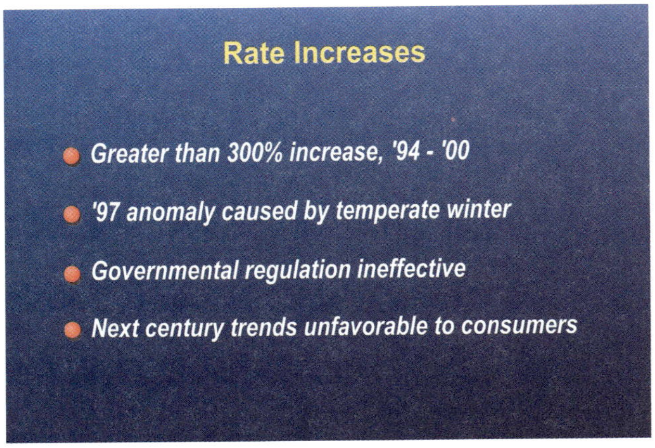


figure 5–13a

Right Screen

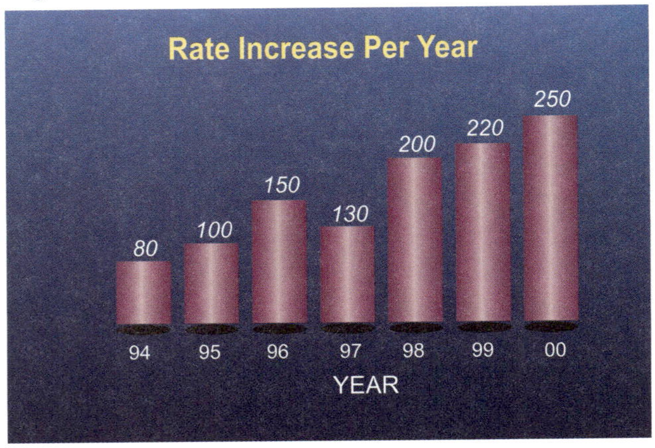


figure 5–13b

Figure 5–13a presents bullet point statements to explain the bar graph in Figure 5–13b. When viewed side-by-side, a great deal of information is effectively presented to the audience.

Left Screen

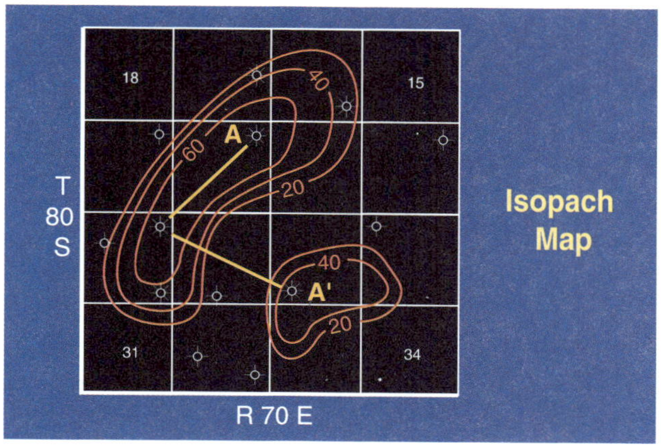


figure 5–14a

Right Screen

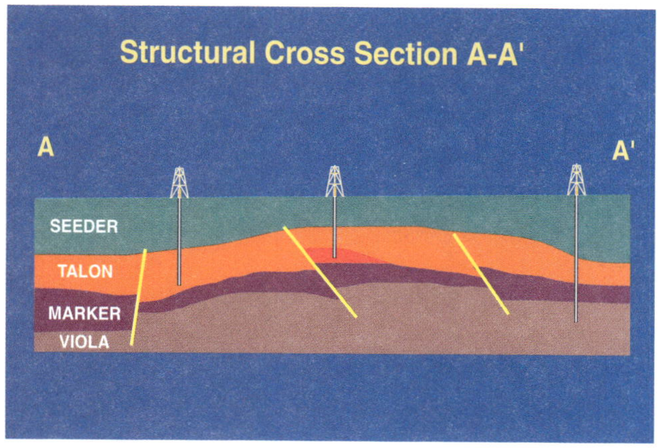


figure 5–14b

This pair of slides presents a map view and a cross-sectional view of an oil field. Displayed side-by-side, the illustrations present the three-dimensional geometry of the geologic structure. Note that the line of cross-section (A-A') is shown on the map.

## *Reduce Excessive Annotation*

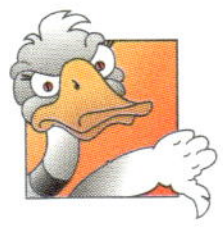

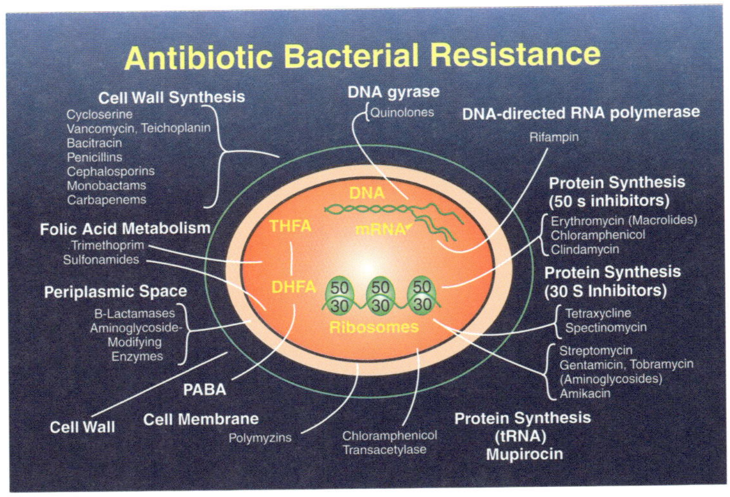


figure 5–15a

An audience simply will not be able to read all of the annotation on Figure 5–15a, and it is difficult to fit it all in using appropriate font sizes. If the audience can't read it, why include it? The redrafted version (Figure 5–15b) is much more effective. If more information is necessary, draft another slide highlighting some of the detailed annotation. Figure 5–3 provides another example of this principle.

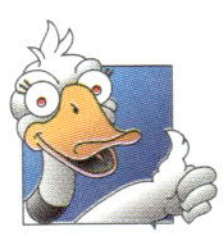

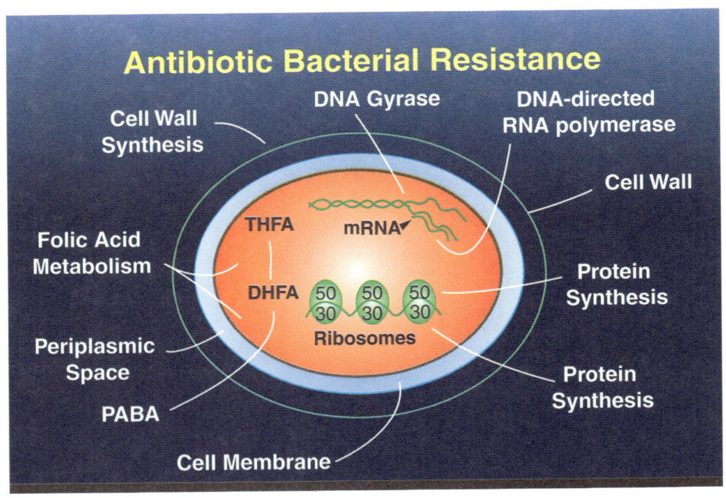


figure 5–15b

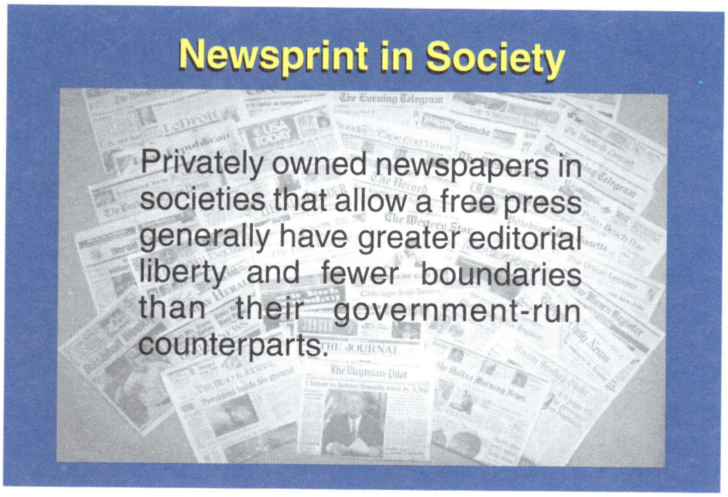


figure 5–16a

*Background Photos can be Useful Embellishments*

The image of newsprint embellishes this word slide. The two figures illustrate different, but equally effective, uses of background photos while maintaining the legibility of the text.

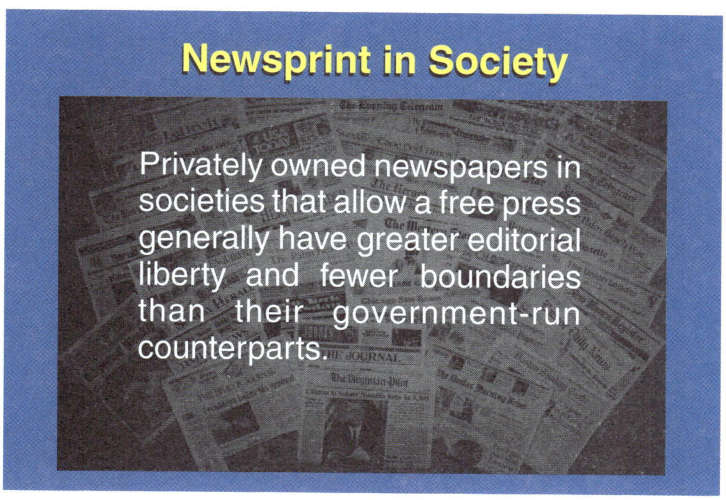


figure 5–16b

## *Use Legible Direct Labeling on Photographs*

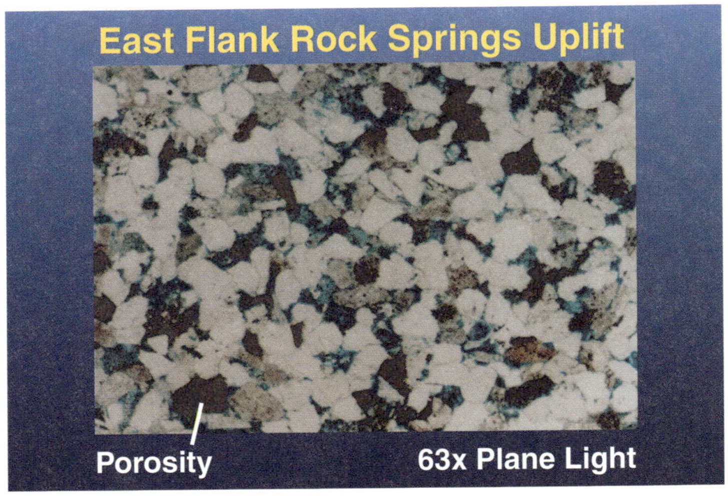


figure 5–17a

This example shows a digital photograph embedded in a slide. Placing annotation directly on the photo allows space to enlarge the photo, but be sure text does not get lost in the background. Notice that direct labeling allows the photo in Figure 5–17b to be larger than in 5–17a. Remember, make sure that the photo fills at least one-fourth of the slide.

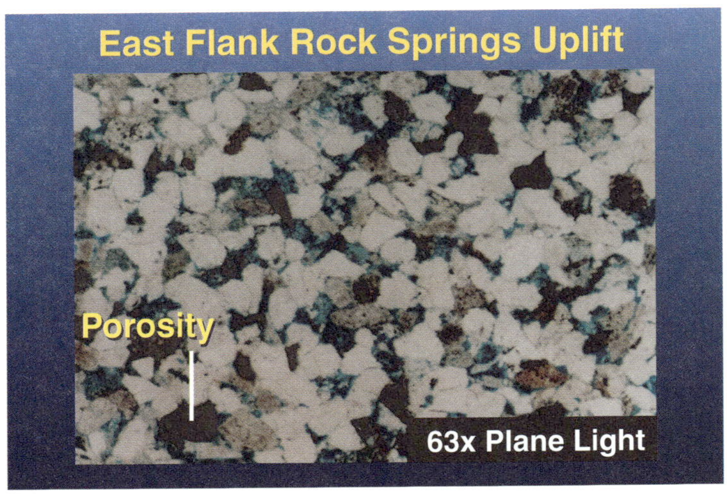


figure 5–17b

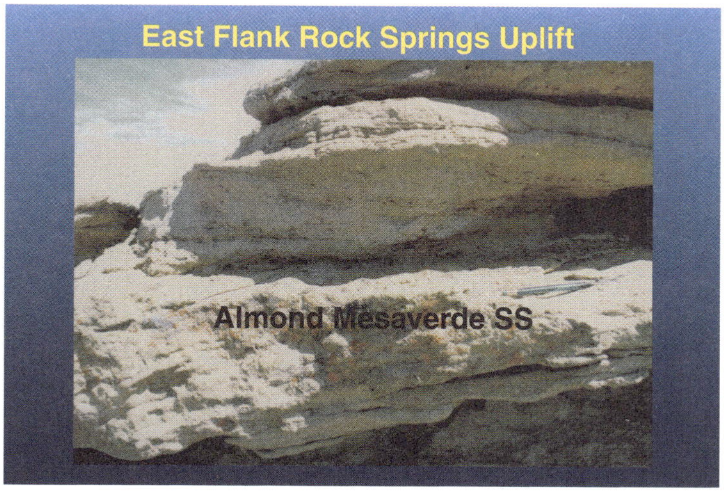


figure 5–18a

The direct labeling in Figure 5–18a is difficult to read because the annotation is lost in the background of the photo. In Figure 5–18, legibility is improved by placing the light-colored font on a shadowed area of the photo.

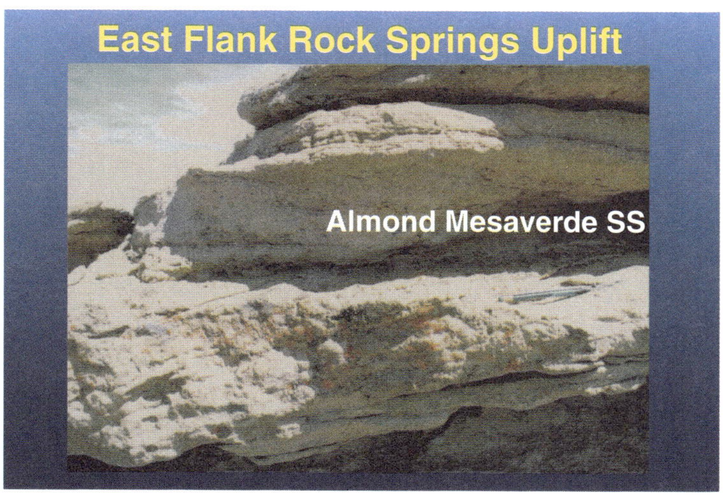


figure 5–18b

## *Use Direct Labeling Rather Than Legends*

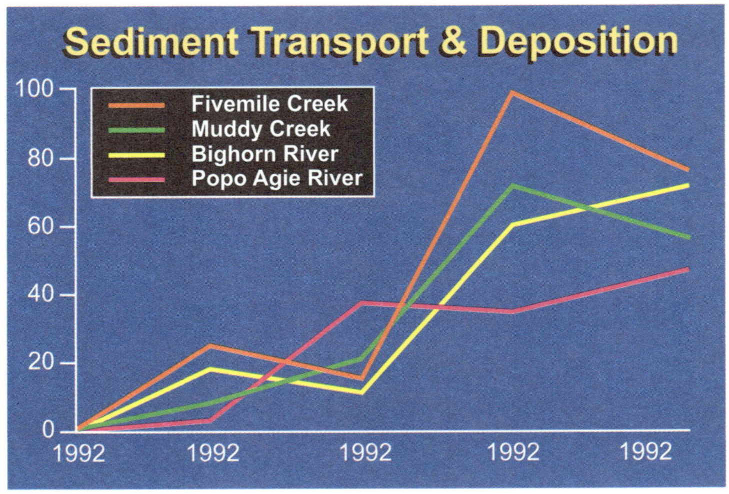


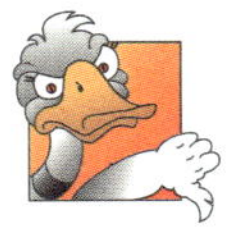

figure 5–19a

Having the lines identified in a legend on Figure 5–19a requires viewers to shift their eyes back and forth across the slide. Direct labeling on Figure 5–19b makes it much easier for the viewers to know which line represents which river. Notice that the annotation is color-coordinated with the data.

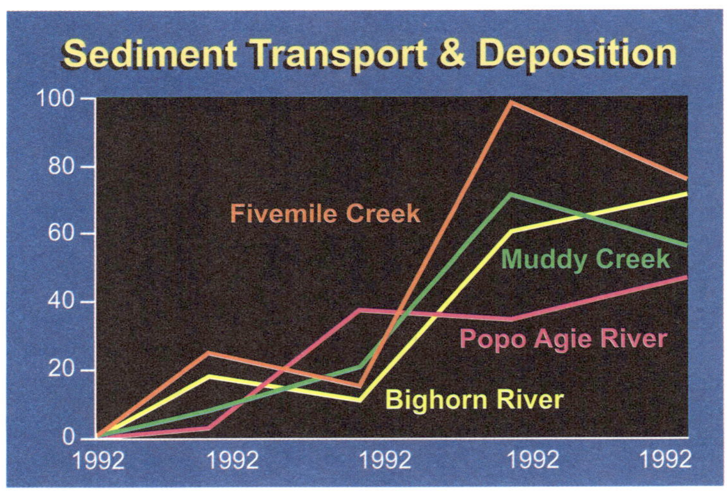


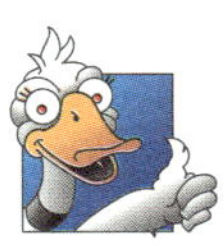

figure 5–19b

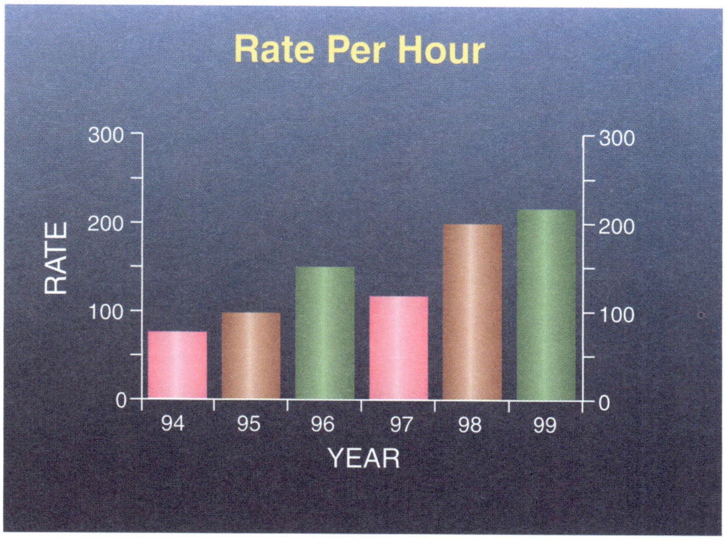


figure 5–20a

## *Replace Framework With Direct Labeling*

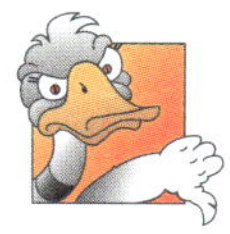

A bar chart with a framework grid is perfectly acceptable if a reader has ample time to determine the values of the bars. However, direct labeling gives the same information at a glance. This also allows the now unnecessary axis lines and labels to be removed.

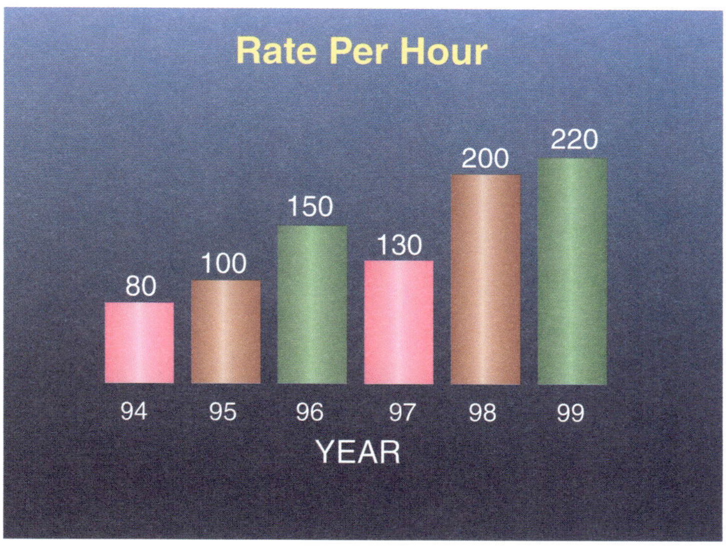


figure 5–20b

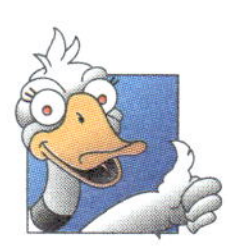

## *Use a Hierarchy of Colors for Titles and Annotation*

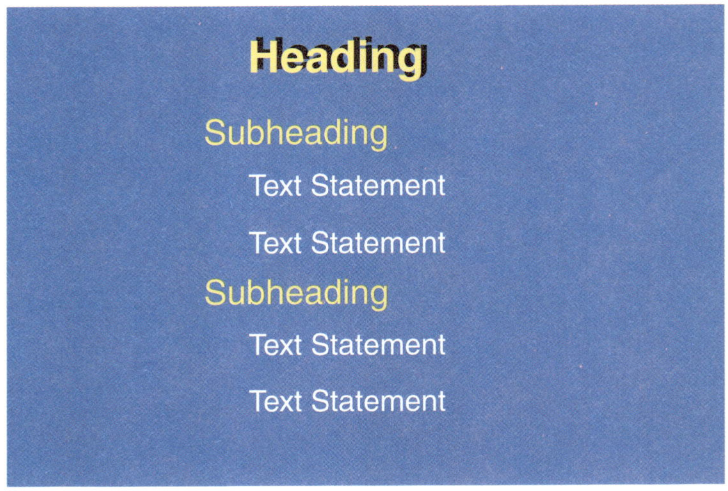


figure 5–21

The two shades of yellow text and the white text create a heirarchy of information, segregate wording, and enhance legibility. Figure 4–1 illustrates this same principle.

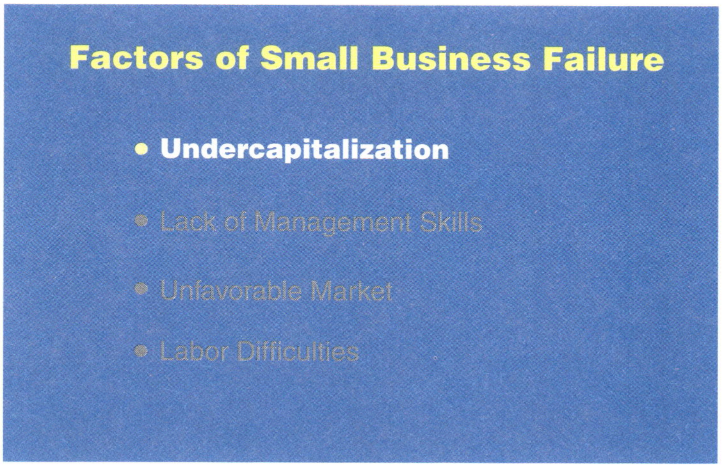


figure 5–22a

## *Use Color to Highlight the Point Under Discussion in a Bulleted List*

In these examples, the first and second bullets are successively highlighted while the speaker discusses those topics.

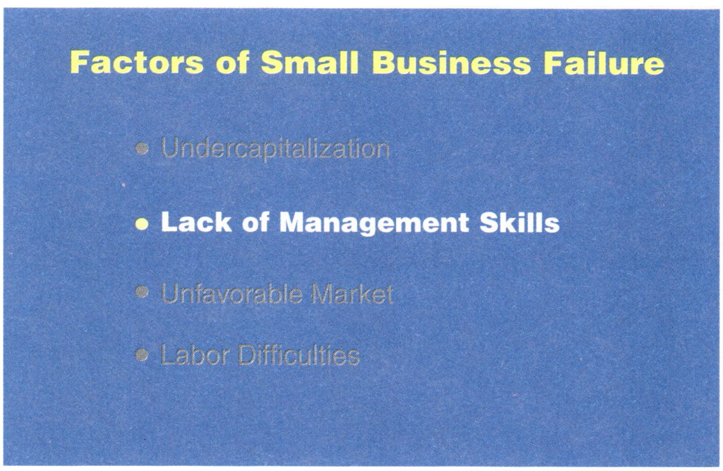


figure 5–22b

## *Adjoin Similar Colors for Gradation/Separate Similar Colors for Contrast*

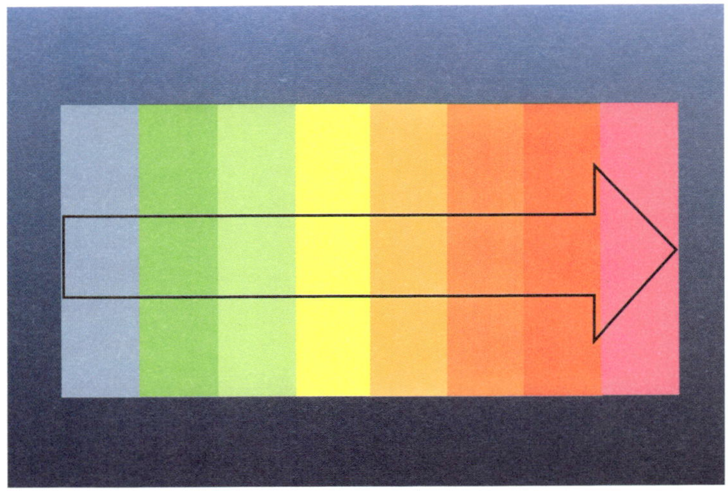

figure 5–23a

Figures 5–23a and 5–23b demonstrate that gradual transitions of color can represent trends or "movement," while adjacent contrasting colors create boundaries and clearly separate graphical elements.

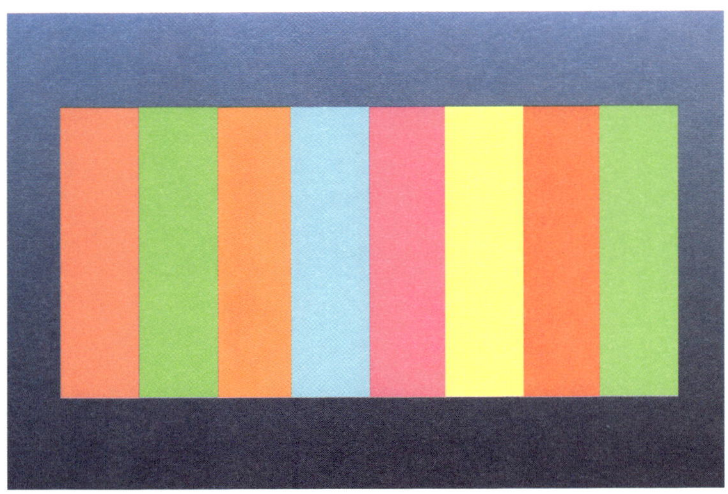

figure 5–23b

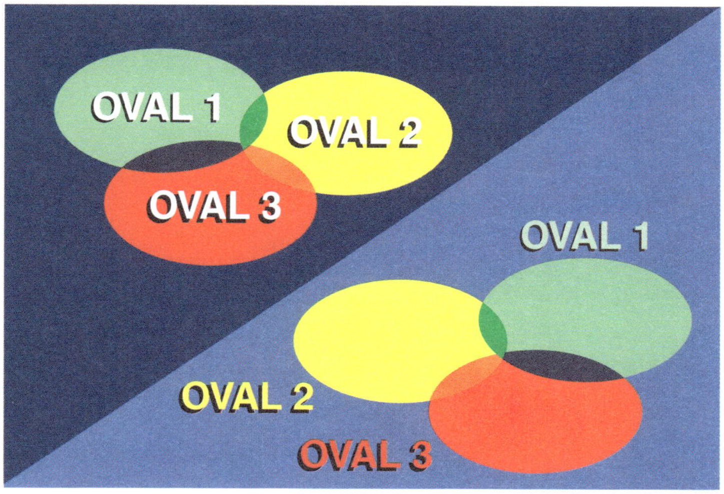


figure 5–24

## *Use Font Colors Consistently*

When labeling objects, use the same text color throughout (upper left), except where labels are color-coordinated with graphical objects to enhance the clarity of labeling (lower right). Look also at the use of color-coordinated annotation in Figures 5–10 and 5–19b.

## *Colored Objects Don't (Necessarily) Need Outlines*

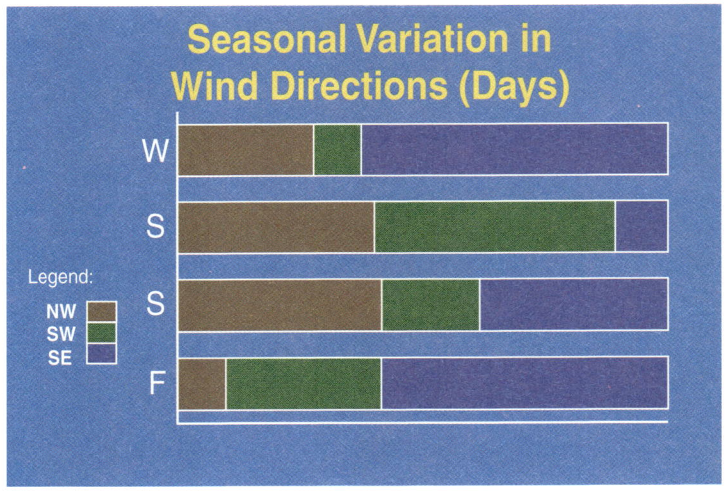


figure 5–25a

Graphical objects with strongly contrasting colors do not require outlines. Save outlines for rare cases where adjacent colors must be similar.

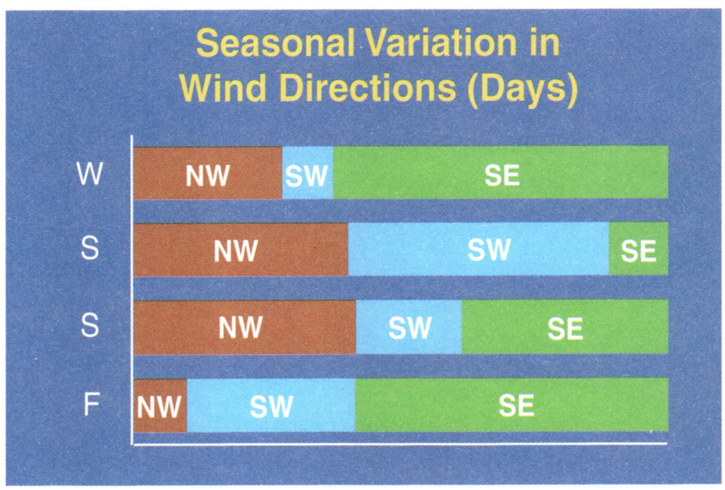


figure 5–25b

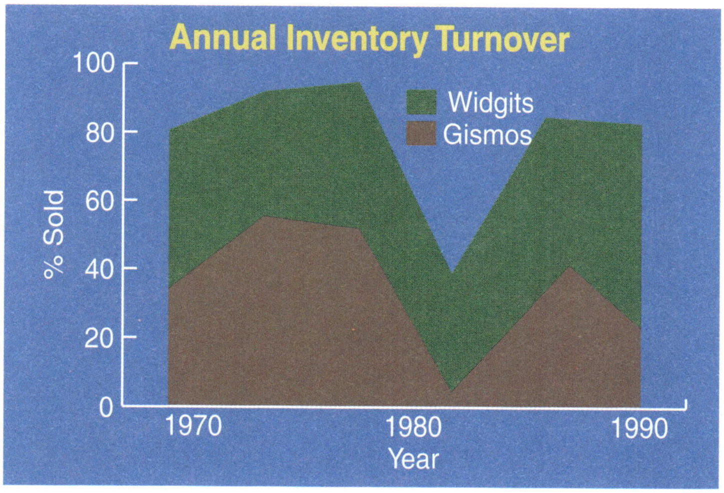


figure 5–26a

## *Use Primary Colors and Standard Fonts*

Primary colors and standard fonts are clearly preferable to cute colors, earthtones, and flowery fonts. Notice also that the annotation "Year" is unnecessary, and direct labeling replaces the legend.

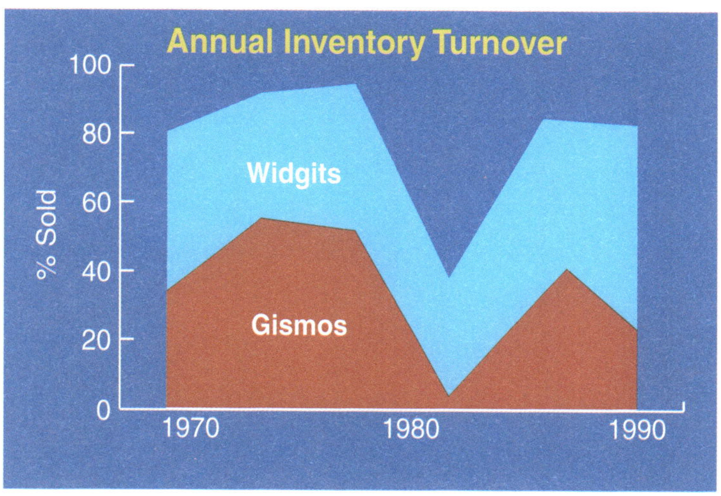


figure 5–26b

## *Replace Symbols With Colors*

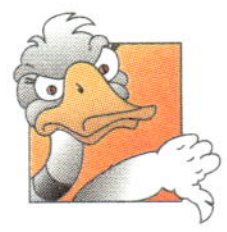

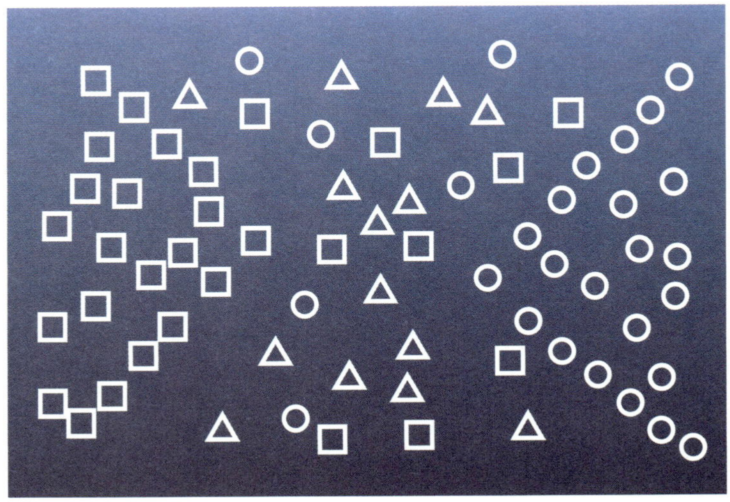

figure 5–27a

The colored dots are much more effective than monochromatic symbols in communicating the distribution of data points.

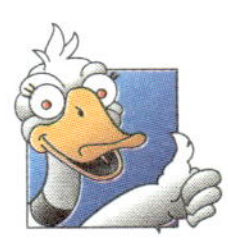

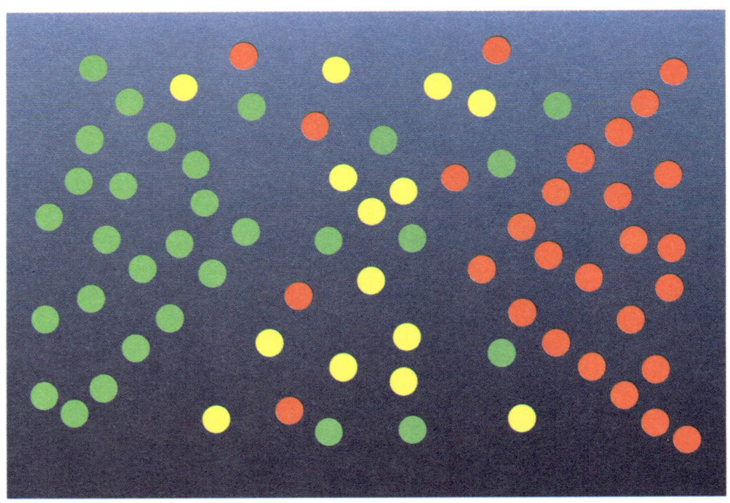

figure 5–27b

## *Replace Line Styles With Colors*

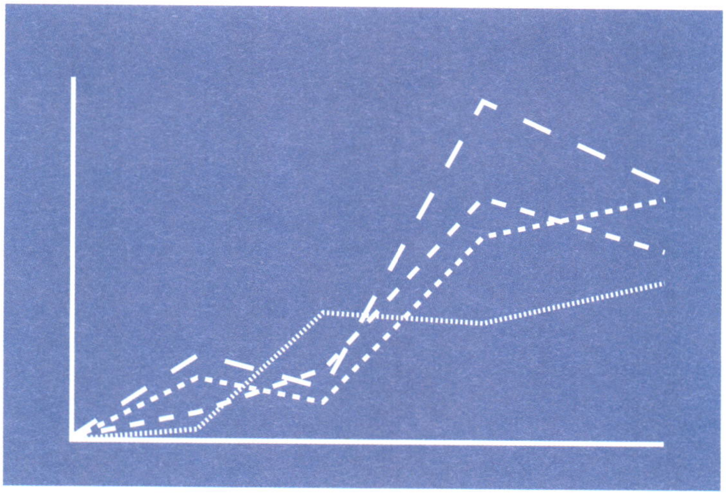

figure 5–28a

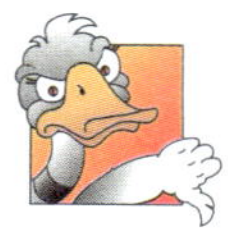

Take a look at Figure 5–28a. How much effort does it take to follow a single line through this crowded clusters of lines? Figure 5–28b demonstrates that contrasting line colors make it much easier to follow individual lines.

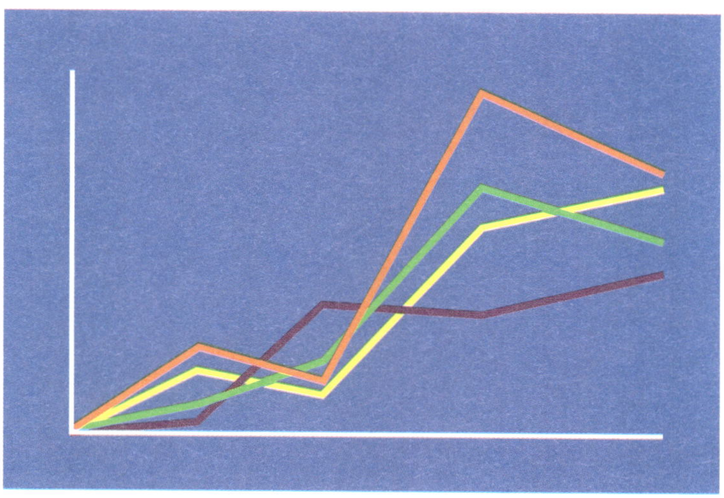

figure 5–28b

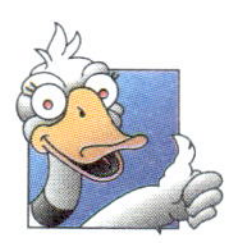

## *Orient Text Consistently for Readability*

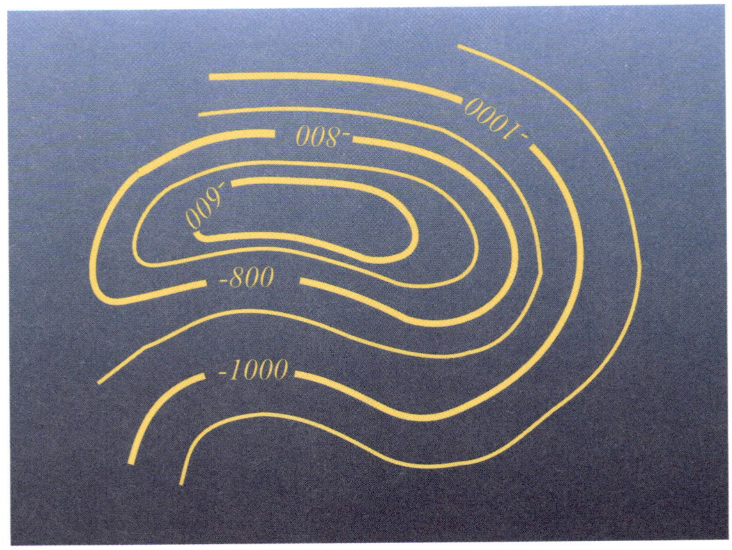


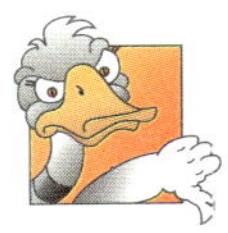

figure 5–29a

Annotations of contour intervals in Figure 5–29a maintain a consistent orientation to the contour lines, resulting in a "flip" in the eyes of the audience. Figure 5–29b fixes this problem and enhances readability.

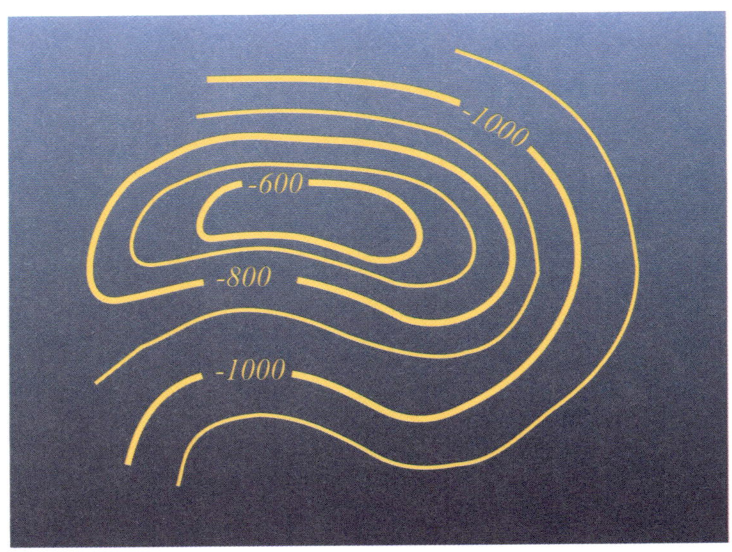


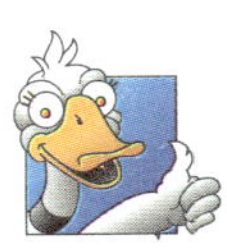

figure 5–29b

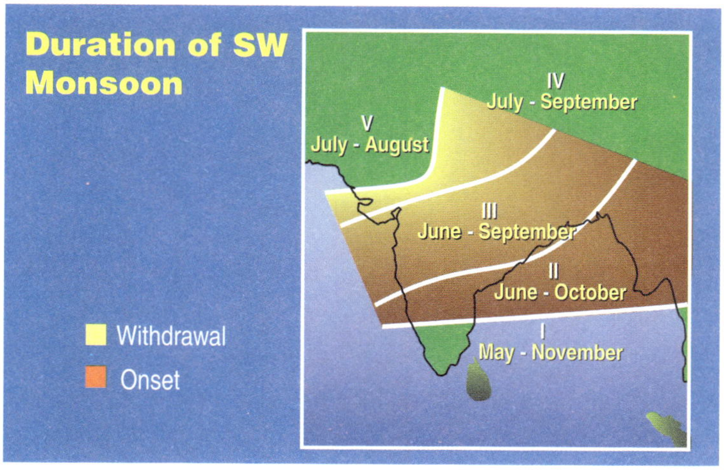


figure 5–30

## *Use Colors to Identify Data Values*

In example 5–30, the color gradient enhances the recognition of monsoon seasons, with dark colors representing early onset – late withdrawal and light colors representing late onset – early withdrawal.

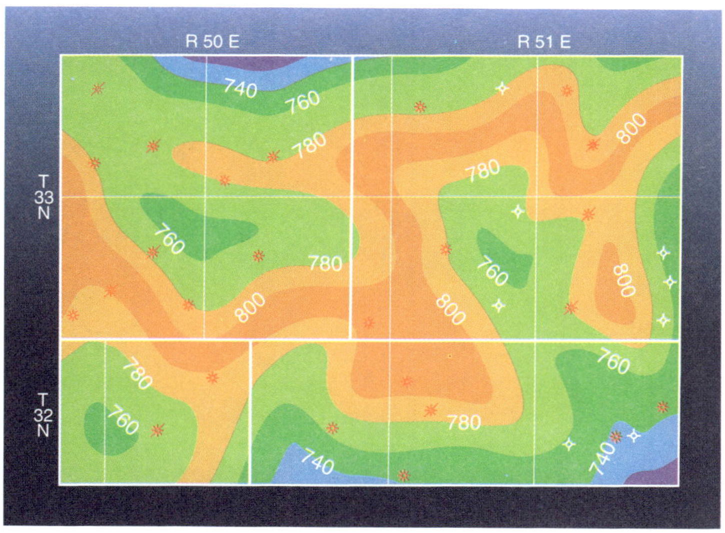


figure 5–31

Figure 5–31 illustrates this technique applied to contour intervals. On this geologic map, purples and blues are low values and oranges are high values. Various other scientific disciplines may have similar color conventions.

## *Use Blow-up to Show Detail*

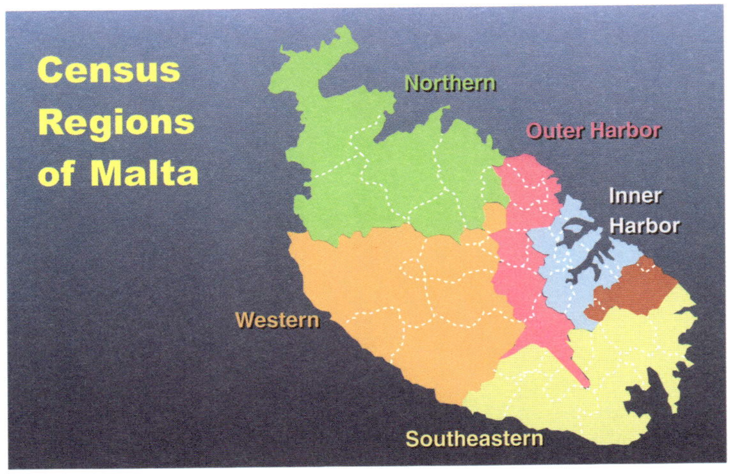


figure 5–32a

These slides present census districts on the island of Malta. On Figure 5–32a, it is not possible to identify individual districts, but it is possible to annotate the names of groupings (regions) of districts. To name the districts, it is necessary to use a blow-up of an individual region (Figure 5–32b). Wherever possible, have the reduced overall image of the map or object on the left side of the slide with the blow up to the right side allowing a natural left to right flow of information.

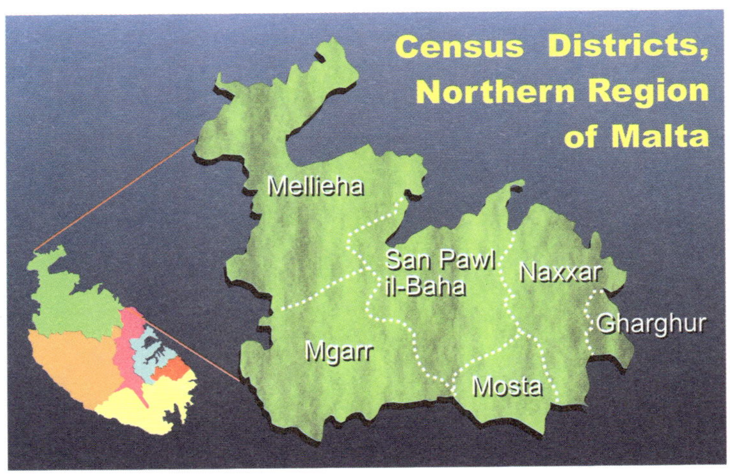


figure 5–32b

# 6

# *Presentations in the Computer Age*

Immediately following the publication of the original *Figuratively Speaking* in 1986, computers began to play a rapidly increasing role in presentation graphics. A set of very powerful tools now exists for illustrating, photo enhancement, and slide design. These advancements have made many of the art preparation techniques in the original book obsolete.

In addition, computer graphics allow dynamic presentations, making use of special effects, in comparison to more static 35-mm slides. We can now add sound, video, and animation to our presentations, as well as many other impressive effects.

## Presentations in the Computer Age

Computers have given us tremendous capability for making presentations. So much capability exists, and it is advancing so rapidly, that it cannot be completely covered in this short guide.

What follows is a description of the general process of drafting effective presentations on a computer. Then we briefly describe the growing and exciting technology for delivering those presentations electronically.

## Overview of Creating Slides on a Computer

### Creating 35-mm slides with a computer

- Select or design appropriate template(s) and page size (71/3 x 11inches for most software)
- Select a background
- For text, use the text tool of the presentation graphics program or import or paste text from a word processor
- For graphics, use the drawing tools of the presentation graphics program and/or import or paste scanned images, clipart, or files from graphics programs
- For photographs, import electronic files from a scanner, digital camera, or PhotoCD
- Compile and lay out these elements into well-designed slides
- Save file in a format for film recording (include fonts with file if using a service bureau)
- Transfer the file to film with a film recorder
- Have film processed into 35-mm slides

Computer software programs greatly facilitate the design and production of slides. They increase the versatility of presenters and have eliminated many of the cumbersome drafting and photographic techniques described in the original *Figuratively Speaking*. Last but not least, these programs have the great advantage of allowing endless iterations of templates, layouts, and designs, including the easy modification of files for different audiences or purposes.

We are talking here primarily about Macintosh® and PC computers, with full-capability graphics programs, digital photography, scanners, and word processors. However, this is not a guide to the use of software programs, and we will not tell you which icons to click or buttons to push to select text or colors or to import graphics or photographs into any particular program. That you will have to learn on your own for the software of your choice. Rather, we can provide an overview to the process of creating professional-quality slides on computers, regardless of the software in use.

Throughout this discussion, remember that the design principles discussed in the previous chapters are fully relevant to anything that you can create on a computer. Presentation graphics programs offer a seemingly endless number of embellishments and special effects. An admonition is appropriate here. Use only those features that enhance the functionality of the slide. Just because your software can do something clever does not mean you should use that feature!

Figure 6–1 provides an overview of the process. Central to the process is a presentation graphics program, such as PowerPoint®. These programs allow the selection or creation of templates and the choice of backgrounds. They have text capability (you can type your text), various degrees of drawing capability (you can do some or all of your artwork), and the ability to import the digital files of text, graphics, tables, and photographs from scanners or other software programs. Any or all of these components, as appropriate, are combined into slides. When ready, electronic files are prepared for imaging on film recorders (i.e., the slides are shot on special devices), and the film is processed into 35-mm slides. Alternatively, the slides are organized into an electronic presentation for display on a computer monitor or projection from a digital projector.

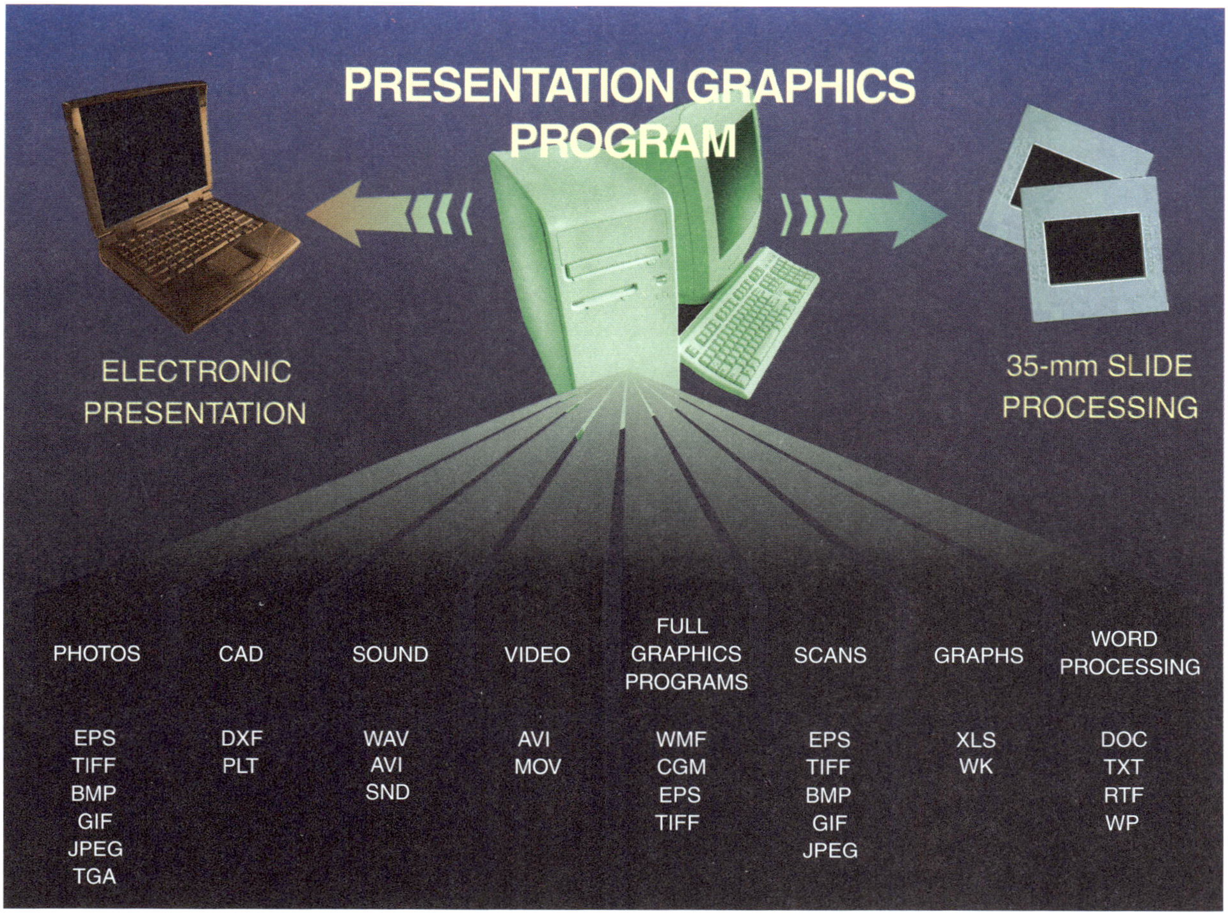


figure 6-1 *Presentation graphics in the computer age involve a wide variety of digital files compiled for output as 35-mm slides or electronic presentations. Each of the acronyms for the digital files is defined in the Glossary (Appendix B).*

Most presentation graphics programs facilitate the process by providing ready-to-use templates. The templates have the correct dimensions and borders. They have text boxes and graphics boxes appropriately placed on the slide. You can select one or more templates for use in a series of slides. For example, one template may be specifically for word slides, and another may be for slides that include graphics. Once you have chosen a template and a background, you are ready to add the content.

An understanding of electronic file formats is important to the process of importing text and graphics into a presentation graphics program. Scanners, software for graphics, charting, and word processing, and digital photography create electronic files in a variety of formats. Figure 6–1 lists several representative formats produced by these programs.

Presentation graphics programs import these files through "filters." The number and types of filters that a program has determines its import capability, and some are more capable than others. For example, if the presentation designer needs to import an engineering drawing from a computer-assisted drafting (CAD) program, the presentation graphics program in use must be able to import a DXF or PLT file. If not, the artist will have to take another approach, such as scanning a hard copy print of the drawing, saving the scan in a format such as TIFF, and then importing that file.

In addition to import capability, Macintosh® and Windows®-based presentation graphics programs have copy and paste capability. Text, charts, graphs, and other graphics may be copied from another application and pasted into a slide. For example, a table created in a spreadsheet program may be copied and pasted into a slide.

Once the design and layout of the slides are complete, the file is saved in a format appropriate for slide imaging. Imaging (or shooting) slides from electronic files involves a peripheral device called a film recorder that contains 35-mm slide film. After imaging, the film is removed for developing in the normal manner.

If the computer used for designing the slides has a film recorder as a peripheral device, the recorder will read the files from that computer's hard disk. If not, the files must be copied onto a diskette or other storage device, or perhaps sent by e-mail, to another computer. This will always occur if the files are taken to a service bureau for imaging and processing. In this case, other considerations arise, such as including fonts with the files. We discuss these issues in Appendix A, "Working with a Service Bureau."

## *Software*

Software programs of many types may be employed in the creation of presentation graphics. Presentations vary from simple to complex in artistry and technology. At the high end of that spectrum, a presentation designer may require a wide variety of software features and functionality and employ several programs to achieve the desired presentation.

Software features for creating presentations include:

- Drawing tools for creating original illustrations
- Graphing tools for creating charts and graphs
- Photo editing tools for manipulating and editing digital photographs and scans
- Scanning commands to drive scanner hardware
- Templates for the layout and format of slides
- Presentation software for building electronic presentations
- Animation and special effects for adding advanced visual features to electronic presentations
- Audio or video embedding for adding sound and video to electronic presentations

The presentation software field is very competitive, and many companies have products on the market to perform these functions. Some of these software programs provide only one or a few of the features listed above, while others bundle many or all of them into feature-rich packages. Some are more sophisticated than others in how they handle these functions.

**Software for PC and Macintosh computers commonly used in presentation graphics**

- PowerPoint®
- Canvas®
- Freehand®
- Photoshop®
- CorelDraw®
- AutoCAD®
- Illustrator®
- Director®

## Technical Fundamentals of Computer Graphics

Technical topics that are fundamental to understanding graphics and presentation software are:

- Raster versus vector images
- Digital file formats
- Color formats
- Image resolution

### Raster Versus Vector Images

Graphics programs create images of two basic types: raster images and vector images (Figures 6–2).

Raster images, also called bitmapped images, comprise groupings of squares called pixels. Pixels form a grid of squares across the area of an image, and individual pixels have the properties of size, position, and color. The size of pixels is constant across an image and is expressed as dots per inch (dpi). For example, the centerpoints of pixels in a 300-dpi raster image are one three-hundredths of an inch apart horizontally and vertically. Each pixel is assigned a solid color (including black and white), and the distribution of pixels of different colors creates the image.

The size of the pixels (dpi) defines the resolution of a raster image. For presentation graphics, resolution typically ranges from 72 to 300 dpi. When you zoom in on a bitmap or if you enlarge the image, individual pixels become evident and appear stair-stepped (jagged) and unclear. The resolution required for a sharp, clear image depends on several factors, such as the means of projection (35-mm slide, computer monitor, digital projector, etc.), the amount of detail in a drawing or photo, and the use of gradational colors and grayscale shading. Often, experience gained from trial and error is necessary to estimate resolution requirements.

Note that it is possible to scale down the resolution of an image, but it is not possible to scale up the resolution. That is, a 300-dpi image is easily resampled to a 72-dpi image, but a 72-dpi image should not be resampled to 300 dpi. Resampling to increase resolution creates a larger digital file without increasing the sharpness of the image.

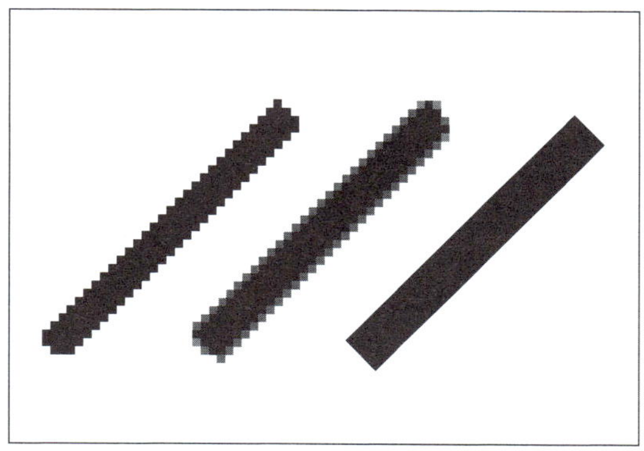

figure 6-2 : *Computer generated lines. Left: a bitmapped line. Center: a bitmapped line with anti-aliasing. Right: a vector-based line.*

Photo editing programs have a function called anti-aliasing that acts to smooth the stair-stepped edges of bitmaps (Figure 6–2). This feature adds light shadings of color around the edges of a bitmapped image, achieving a blended effect that sharpens the appearance of text, lines, and solid fills when viewed without zooming.

Vector images, also called object-oriented or draw images, comprise a series of points connected by lines. They are created mathematically within the software program. Graphical vector images are called "objects," and have the properties of shape, size, position, and color. Objects that outline a closed area have an additional property that defines the color (fill) of that area. On most programs, vector images lack the jagged, pixelated appearance of raster images and retain smooth outlines even at very high magnifications. The color of an object or filled area can be solid or gradational.

### Electronic File Formats

A variety of digital file formats (Figure 6–1) are important in presentation graphics. The primary distinction between these formats is whether they are raster or vector files. "Native" formats are specific to individual software programs, and "standard" graphical file formats are created for import and export by many different programs.

"Standard" graphical file formats that are common in presentation graphics are:

- BMP – "Bitmap," a raster format
- TIFF – "Tagged Information File Format," a raster format
- EPS – "Encapsulated Postscript," supporting vector and raster formats
- CGM – "Computer Graphics Metafile," a vector format
- WMF – "Windows Metafile," a vector format

Examples of "native" file formats include PPT in PowerPoint, AI in Adobe Illustrator, and DWG in AutoCAD.

Files imported from word processing and spreadsheet software are also important. These formats include DOC (Microsoft Word), WP (WordPerfect), XLS (Microsoft Excel), and WK (Lotus 1–2–3).

Import filters allow software programs to bring in (import) files of various standard formats and save them in the native format. For example, CorelDraw can import a CGM file and save it in its native format, CDR. Similarly, export filters of a program allow it to save the native file format in one of the standard formats (CDR to CGM).

### Color Formats

Important color formats, including black and white, are:

- One-bit black and white—The pallette consists of two values, solid black and solid white.
- Grayscale—The pallette consists of 256 shades of gray, ranging from 0 (pure black) to 255 (pure white).
- Indexed color—The pallette consists of 256 discrete colors.
- RGB—Colors are built of a mix of red, green, and blue, yielding a pallette of millions of colors.
- CMYK—Colors are built of a mix of cyan, magenta, yellow, and black, creating a pallette of millions of colors.

This list is arranged in order of increasing information content in the color choices. The list is also arranged in order of increasing file size. That is, for any one graphic, CMYK colors will create a larger file than RGB colors, which will in turn create a larger file than indexed color, and so forth. The top end,

CMYK does not yield the greatest color and clarity on your monitor. RGB is actually a larger color space incorporating more of the visual spectrum than CMYK. CMYK is really only necessary when going to pre-press. Computer displays and slide imaging cameras operate with RGB colors. Therefore, presentation designers work with one-bit black and white, grayscale, and RGB color formats.

### File Size

The storage space occupied by digital files for presentation graphics is expressed in kilobytes (kb) or megabytes (mb). File size can be influenced by the resolution (dpi) of a raster image, the number of objects in a vector image, color format (described earlier), and file format. File size increases with increasing resolution of raster images and with increasing number of objects in vector images. As an example of the influence of file formats, the file size of an image saved in TIFF format will generally be larger than the file size of the same image saved in JPEG format. Table 6–1 illustrates the effect of file formats and color formats on the file size of a sample 300-dpi image. This example describes a test done in a single software, and other applications may yield different file sizes.

File size is highly important in presentation graphics because of hardware limitations (memory, hard drive space, and storage media) and the speed of programs. In general, presentation designers make file sizes as small as they can without losing any of the quality or features of a presentation.

Techniques for managing file size are:

- Use the lowest resolution of images feasible while maintaining the clarity of the image.
- Choose a file format that yields the best quality image with the smallest file size.
- Convert CMYK images to RGB or indexed colors. (Compare the image quality between RGB and indexed color when making this choice.)
- Limit the number of vector objects in a drawing. (This is most critical with files imported from some CAD-type programs, which treat lines as a series of separate segments, each of which is an object, and build letters and symbols from several separate elements.)

### Limitations Imposed by Display Media

The method chosen for displaying presentations has a large effect on the projected resolution of images. Regardless of the resolution at which an image is created, the projected resolution of that image can differ on a computer monitor, a digital projector, or 35-mm slides.

The number of pixels displayed horizontally and vertically determines the resolution of a computer monitor. Video cards

Table 6-1 *Effect of file format and color format on the file size of a 300-dpi image*

| | CMYK | RGB | INDEX COLOR |
|---|---|---|---|
| TIFF | 3.35 mb | 2.63 mb | 782 kb |
| JPEG | 752 kb | 413 kb | – |
| EPS | 21.5 mb | 17.2 mb | 17.2 mb |
| BMP | – | 6.24 mb | 2.08 mb |

**A caution about JPEG file format:** JPEG uses a lossy compression that loses information each time it is opened and re-saved. JPEG should only be used for the final version of an image, never on a working version. It is great for foot transmission and saving disk space, but be careful!

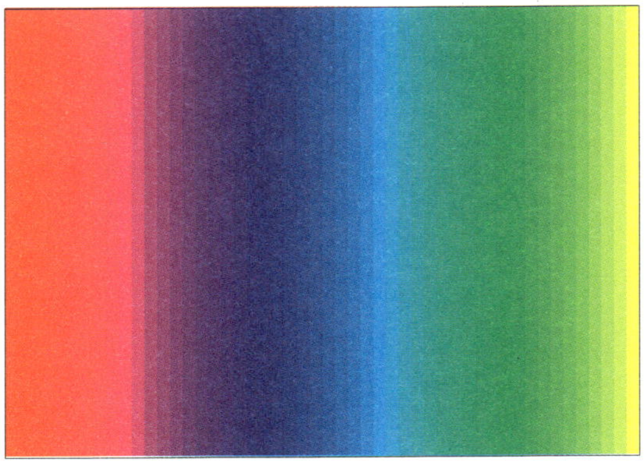

figure 6-3a

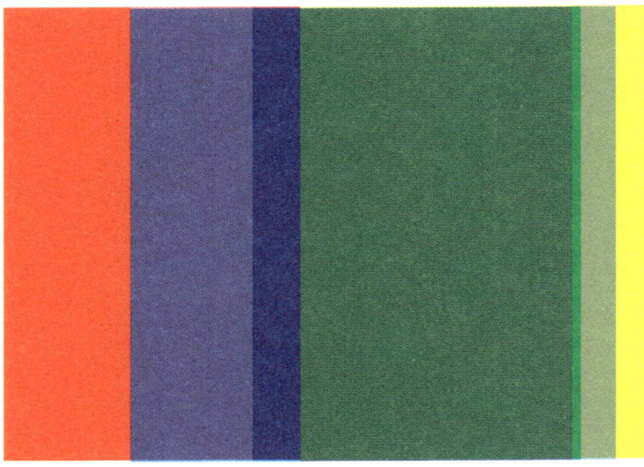

figure 6-3b

figure 6-3 *The color-support settings of a monitor determine its ability to display color gradations. (a) Monitor settings that support RGB colors (millions of colors) yield smooth color gradations. (b) Lesser color support, including indexed color (256 colors), yields abrupt color bands and cannot display smooth gradients.*

control this resolution, and most of these offer a choice of settings. A low-resolution setting displays 640 pixels horizontally and 480 pixels vertically, for a total of 307,200 pixels. Other standard resolutions are 800 x 600 (480,000 total pixels), 1024 x 768 (786,432 total pixels), and higher. All of these settings are low resolution in comparison to the higher resolution available in print media and 35-mm slides.

An artist can choose both monitor resolution and image resolution, both of which affect the clarity of an image. Image clarity decreases with lower image resolution and lower monitor resolution.

The number of colors supported by a moniitor can lead to dramatic differences between the way colors are designed to appear and how they actually look on the monitor. This issue also arises with the use of digital projectors.

Figures 6–3 illustrates smoothly blended color gradations and abrupt color transitions. This difference reflects a monitor supporting RGB colors (millions of colors) versus a monitor limited to indexed color (256 colors). The lesser color support cannot smoothly display blends and color gradients.

Digital projectors, including liquid crystal display (LCD) projectors, project images directly from digital files onto viewing screens (Figure 6–4). These devices have a similar range of resolution to that of computers. When projected, an image becomes very large (screens in some venues may be 20 ft high), and the pixel size increases proportionately. This causes low-resolution images to lose sharpness and clarity.

35-mm slides offer an advantage over digital projectors in the preservation of fine detail on projected images. The process of exposing digital files to slide film involves hardware peripherals called film recorders. The resolution of a slide is measured in lines, with 4000 lines being standard. Lines are horizontal and contain the color information that builds the image. The large number of lines will capture even the highest resolution achievable in computer graphics.

## Calibrating A Monitor to An Output Device

One of the great challenges of computer graphics is controlling the color of a slide, print, or plot. "What you see is what you get" (WYSIWYG) only applies to position and layout; it definitely does not apply to color. Most often, your output will not match what you saw on your monitor.

Several factors contribute to this problem. Every output device, including printers, plotters, and film recorders, has its own color characteristics. Even individual printers of the same manufacturer and model will exhibit slightly different color output. The various software packages do not even define standard colors with the same color values. Programs have one set of color values to create the screen representation and another set of values to define that color for printing. Finally, most monitors are not calibrated in any way at all.

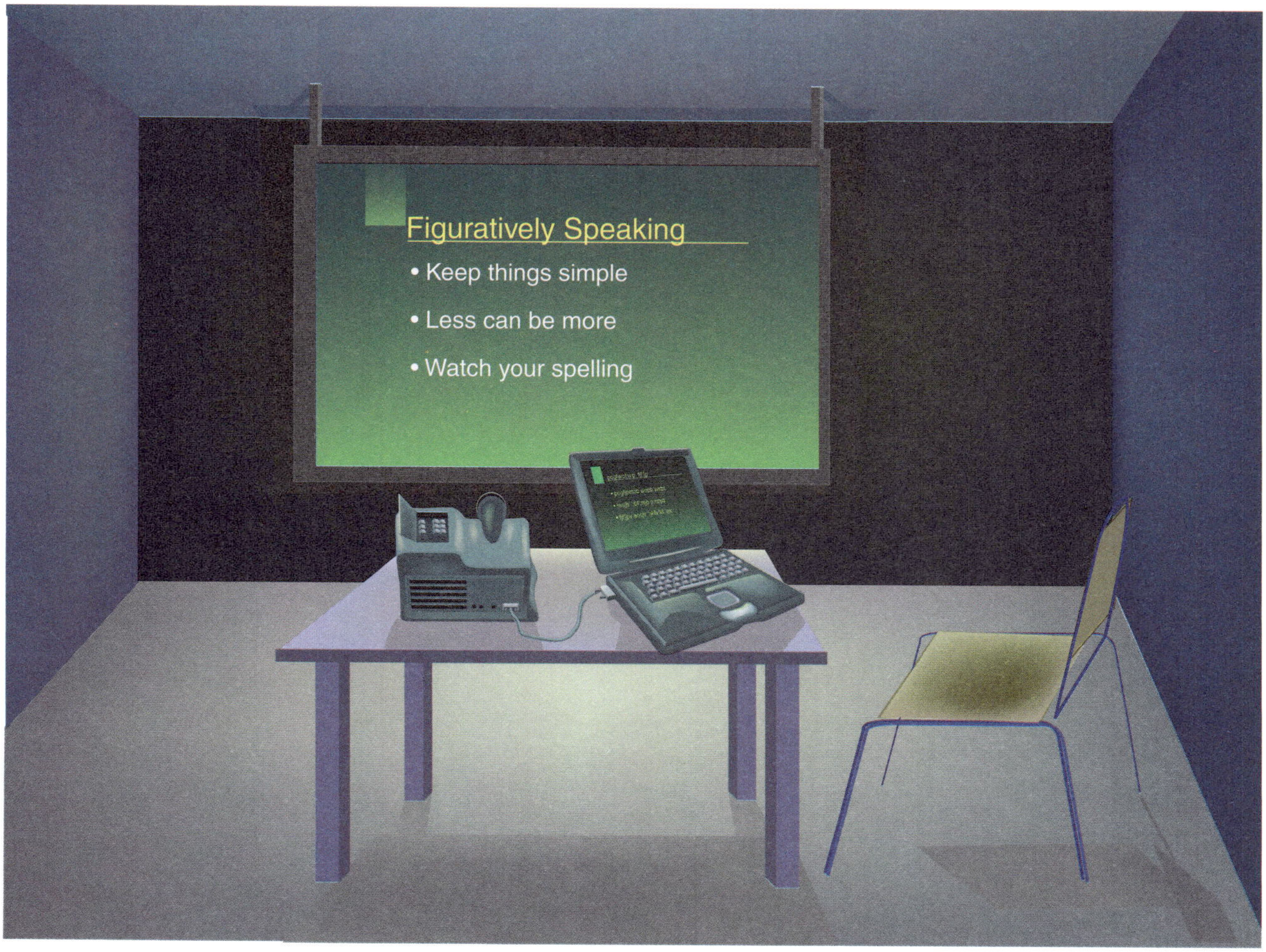


figure 6-4 : *A digital projector running off a laptop computer projecting electronic files onto a screen.*

Without going deeply into color theory, here is a way for you to gain some control over the appearance of your final output.

- Create a file that uses the palette or set of colors that you tend to use most often or are planning to use for a new presentation. This could be several blocks of color on a few pages or simply a photo that you need to be sure looks as good as possible.
- Send that file to your service bureau and have them run it on your chosen output device (color prints, slides, etc.). A good service bureau will be happy to test files for you so that the final products will meet your expectations.
- When you receive the output, open those same files on your computer and look at the differences between the screen and the product. Using the controls on the monitor, if available, or the system-level software controls, adjust the colors of your monitor to match the appearance of the test output as closely as possible. Some colors will never match from print to screen, so look for the best overall balance of colors.

This simple system will be very helpful if done carefully and will help avoid that all too common refrain, "Why doesn't this print look like what I saw on my monitor?"

## Electronic Presentations

The original *Figuratively Speaking* is commonly referred to as a "slide manual," and slides are still in common use. However, because computers allow us to skip the process of shooting 35-mm slides on film, if we wish, and deliver our presentations directly from the computer, we have to loosen our definition of a "slide" to include presentations on, or projected from, computer monitors and digital projectors. These "slides" are not necessarily static; rather, they may be fluid, animated, and enhanced with sound or video (Figure 6–5).

Various devices allow the projection of presentations to audiences directly from computers, the most fundamental being the computer monitor itself. An electronic presentation might involve several people gathered around a 15-inch monitor, the use of a larger monitor in front of a seated audience, or the projection of images through a digital projector onto a screen in a large meeting room. The technology that enables electronic presentations is evolving rapidly.

Electronic presentations involve the same principles of slide design and techniques for making effective presentations that have been discussed earlier in this book. However, the design of "slides" for electronic presentations is slightly different from the design of 35-mm slides. Specifically, these differences are:

- Whereas 35-mm slides have 2– x 3–inch dimensions, computer monitors are more square, with 3– x 4–inch dimensions, requiring slightly different layout templates (Figure 6–6).
- Electronic presentations can incorporate a wide variety of special effects (transitions, fades, voiceover, embedded video, animations, etc.).

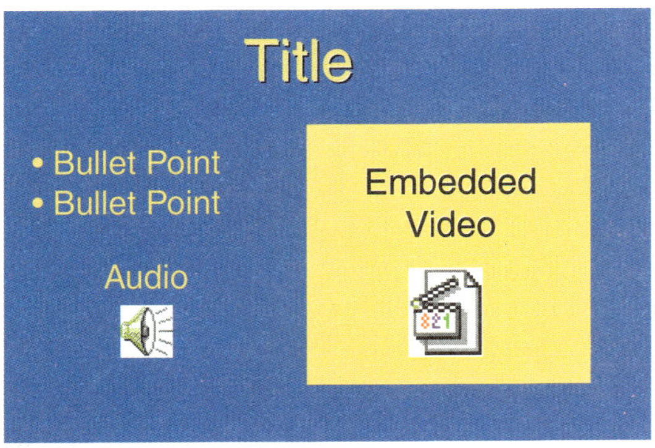


figure 6-5 : *Layout of a slide with embedded video and sound.*

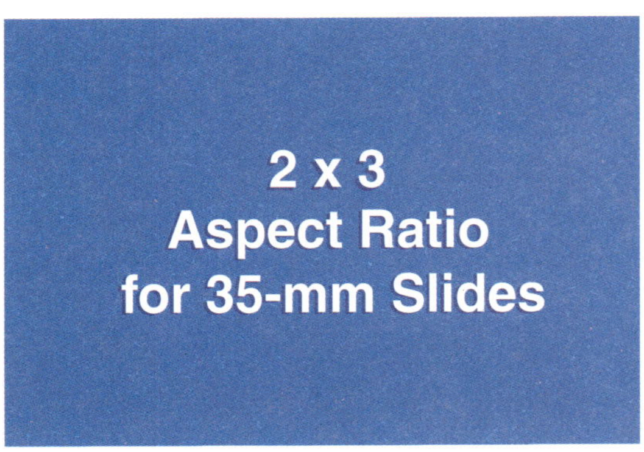


figure 6-6a

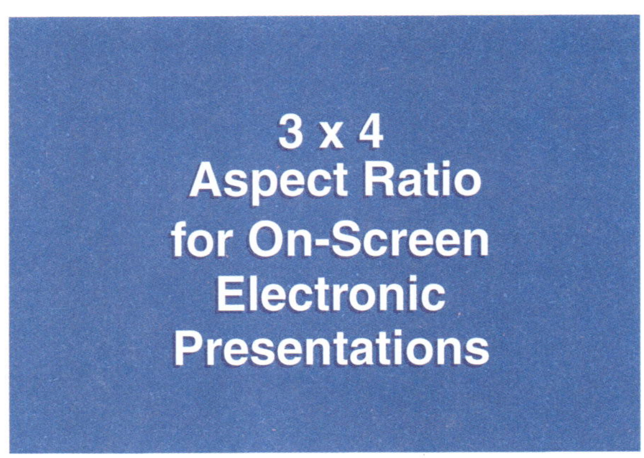


figure 6-6b

figure 6-6 : *The dimensions of slide templates for 35-mm slides are different than those for on-screen, electronic presentations.*

### Special Effects

Special effects are powerful tools for adding information and dramatic effect to presentations. These include:

Transition effects: Fades, builds, fly-ins, and other effects that allow the addition or deletion of information in a slide in an eye-catching manner and take the place of advancing 35-mm slides in a tray.

Animation: Logos, graphics, and text can be animated, either purely for embellishment or for practical purposes such as showing the workings of a piece of equipment, a geologic event, or the unzipping of DNA during cellular division.

Data points or other features that require special attention can blink or flash. This is especially useful for highlighting single points in crowded data fields. If a feature is added to a slide, blinking it two or three times will help the audience notice its presence.

Sound: The addition of voice, music, or other sound effects (e.g., the squeal of tires in a presentation of crash test data) is possible by embedding a sound file in the presentation.

Video: Adding video of a speaker ("talking head"), test procedure, piece of machinery, or any other pertinent action is possible by embedding a movie file in the presentation.

### Using Digital Projection Equipment

Presenters are increasingly making use of digital projectors, instead of 35-mm slides, to project their presentations onto a screen. In fact, 35-mm slides and slide projectors are rapidly becoming yesterday's technology. With this change, the days of burnt-out projector bulbs, out-of-order and upside-down slides, and slides that won't drop are passing. However, those common problems are being replaced by the possibility of computer glitches, software incompatibility, and corrupt files. Therefore, presenters face a new list of considerations to ensure that the presentation is designed correctly and takes place without technical problems.

For an electronic presentation, a presenter may:

- Bring only the electronic presentation file to the venue and use a venue-provided computer and digital projector
- Bring the presentation file on a laptop computer and use a venue-provided digital projector
- Bring all of the necessary equipment
- Bring an electronic presentation and a back-up set of 35-mm slides

A fundamental caution is that the computer system on which a presentation is given may not yield the same quality of presentation as did the computer system on which that presentation was designed. For a presenter who intends to use venue-provided computers and digital projectors, this caution is particularly important.

A different computer processing speed can either speed or slow transition effects and the pace at which programmed slide changes occur. Software incompatibility, including different versions of the same software, can cause the loss of formatting or, at worst, prevent a presentation from projecting at all. Low resolution (640 x 480) and limited color support (256 colors) of the digital projector can cause lost colors, badly pixelated color blends, and lower overall image clarity.

Steps to prevent problems such as these include:

- Find out in advance the type of hardware and software that will be provided at the venue. Is it Macintosh or IBM-compatible? Will the software project your file format? Can you bring your presentation on a Zip® cartridge, Jaz® cartridge, or CD ROM? Can you connect your laptop to their digital projector?
- Find out the resolution of the projector and the range of colors that are supported by the digital projector. If the resolution and color range are at the low end, reduce the amount of fine detail and the use of color blends in the slides.
- Test the projection equipment as soon as you arrive at the venue. Because you have brought your presentation software (see below), last minute adjustments are possible. If your presentation includes audio, test the audio volume as well.

A second caution is that computers and digital files can fail. Murphy's Law can prevail with presentation equipment as with any other machinery. The computer may lock up. A battery on a laptop computer might run low. Your file may have mysteriously become corrupt and unusable. The solutions to these potential problems lie in back-ups and redundancies.

- Carry two copies of your digital presentation. For example, carry one copy on the hard drive of your laptop computer and a back-up copy on a diskette.
- Have a spare battery and a spare power cord (with extension) for your laptop.
- Carry a spare copy of the presentation software.

Don't forget that a set of slides in a tray is a great backup to an electronic presentation. You can almost always find a slide projector at any venue.

Finally, knowledge of how to organize digital files, including fonts, video files, and audio files, is important.

- Embed your fonts in the presentation or load them all onto the computer to be used in the presentation. If the fonts are not present, the computer will substitute other fonts, possibly causing text to reflow inappropriately.

- Audio and video files must reside in the directory path that contains the presentation file.

The use of digital projection equipment offers several benefits that are worth mention. The speaker may be able to set a laptop computer on or in front of the lectern and, therefore, be able to see what is projected onto the screen while still facing the audience. Cursers and special effects such as blinking highlights can replace light pointers. If a "remote mouse" employing an infrared beam is used to advance the slides, the presenter can control the presentation from anywhere in the room. Finally, the navigation tools of a presentation graphics program can allow the speaker to move quickly forward and backward in the presentation or to bring up supplementary information from other files.

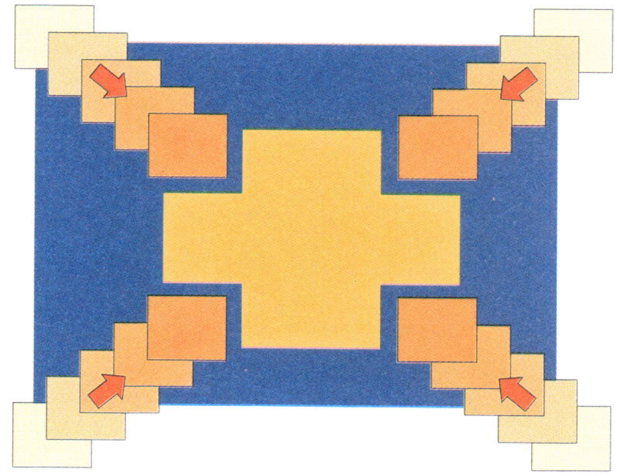

figure 6-7 : *Graphical elements flying into a slide along the shortest, most logical paths.*

## Using Transition Effects

Many presentation software programs offer special effects for making the transition from one slide to the next. For example, these allow the speaker to build slides by adding bullets or new layers, to make objects fly in, or to fade out of one slide and into another. Altogether, these effects can add motion and interest to an otherwise static presentation.

An important caution is to use transition effects consistently throughout a presentation. For example, do not dissolve one slide, peel away the next, and rotate out the third. Do not be tempted to use the entire toolkit of transitions. Doing so will leave the impression that your purpose is to be clever, not to communicate.

If a presentation is divided into a series of topics or sections, use one type of transition effect (e.g., peal aways) to introduce topics and a different type of transition (e.g., fades) to change slides within a topic.

Fly-ins illustrate these principles. In Figure 6–7, graphical elements fly in to the illustration from the most logical directions and across the shortest open space of the slide.

Figures 6–8a and 6–8b illustrate the importance of flying in bullet points from consistent directions (e.g., all from the left or all up from the bottom) and of not bringing the text in across the title, previous bullets, or graphical elements (Figures 6–8c and 6–8d).

Some presenters like to add sound effects to transitions. As an example, a transition effect that shatters a slide and lets the pieces fall to the bottom might be accompanied by the sound of glass shattering. Be careful with this or any other effect that might lessen the professionalism of the presentation.

Experiment with the transitions and other special effects offered by your software. They can liven up a presentation tremendously. Simply keep in mind that the purpose of the presentation is to communicate.

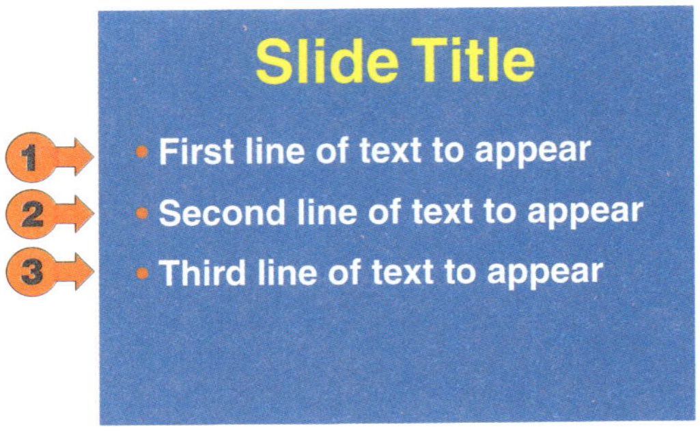


figure 6-8a

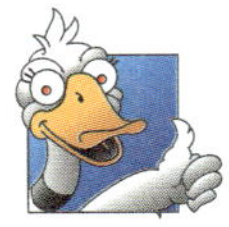

Figure 6-8a & 6-8b :

*Proper ways to "fly in" text onto a slide. In Figures 6-8a and 6-8b, lines of text arrive consistently from the left or upward from the bottom of the slide.*

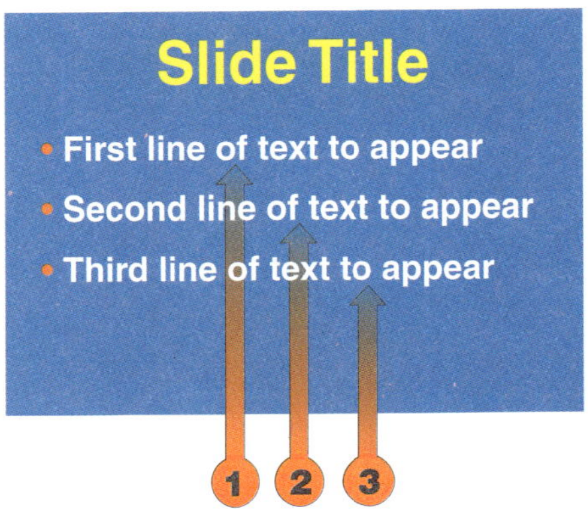


figure 6-8b

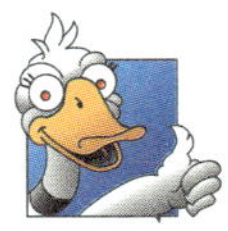

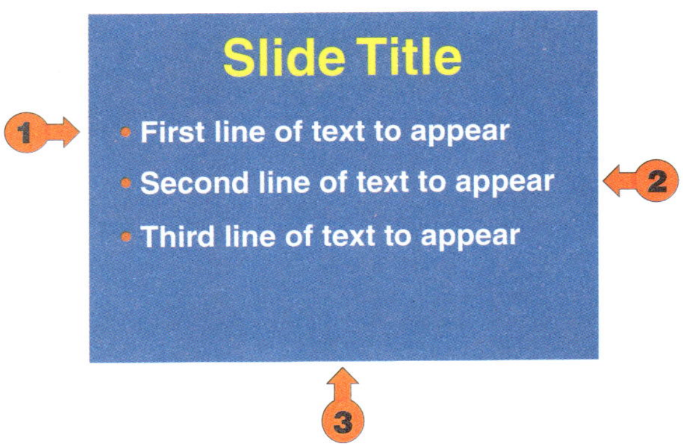


figure 6-8c

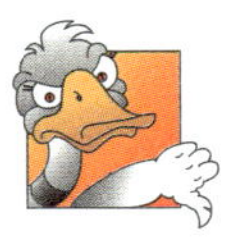

Figure 6-8c & 6-8d :

*Improper ways to "fly in" text onto a slide. Figure 6-8c is improper because the lines of text fly-in inconsistently from various directions, and Figure 6-8d is improper because lines of text cross the slide title and preceeding lines of text as they enter the slide.*

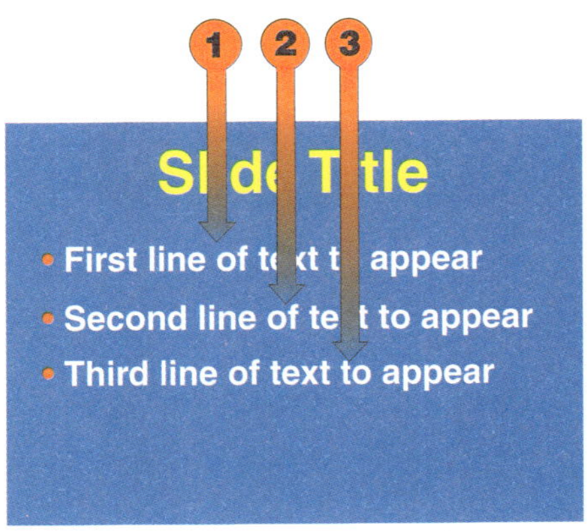


figure 6-8d

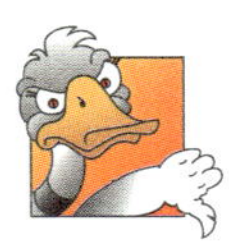

# 7

# *Poster Sessions*

Poster sessions are a popular forum for the presentation of technical information to audiences. The audience can view and read the visual aids at their own pace, and the presenter may be on-hand for discussions and questions. As with slides, the effective presentation of posters requires attention to design criteria, the venue, and the interests of the audience.

Poster sessions provide a forum and format for oral presentations different from that provided by presentations supported by slides. Much like the old teacher-and-blackboard method (except that the "students" may attend at will and are likely to have greater interest in the subject matter), "presenting a poster" allows the speaker to stand in front of or among the audience and discuss the topic with the help of visual aids. In this case, the visual aids are "posters" mounted on a wall or free-standing board. The format also allows the presentation of supplementary materials on tables in front or to the side of the posters, such as samples and specimens, computer programs or videos, and sales material.

In common with oral presentations and published papers, poster presentations benefit from well-written abstracts and introductions, accurate and insightful delivery of the subject matter, and succinct conclusions. Effective poster presentations also require overall graphic design, with effective illustrations organized so that viewers can easily follow the logical progression of ideas and information.

To achieve these purposes, presenters of posters can follow the advice of Chapter 2 and divide their preparatory steps into the sequence:

- Purpose
- Preparation
- Practice

Defining the purpose of the presentation is always the first step, with the necessity of determining the most appropriate information to present to the audience. Preparation can then proceed toward writing an abstract, deciding on the flow of information, and designing and preparing the illustrations. The storyboard technique is a helpful step in this preparation.

Even though prepared remarks may not be required, practice is still necessary because members of the audience may ask to be led step by step through the topic. Invite colleagues to critique your displays and explanations for content, organization, design, and effectiveness. Use their comments to both improve your poster and to anticipate questions from the audience.

## *The Venue*

At poster sessions, authors are provided a board or wall on which they can hang their posters. They may also have tables for additional displays. They are assigned a time to present and a time to pack up. At large conventions, many posters may be presented simultaneously in a specified area of the convention center. Figure 7–1 is an illustration of a forum of this type.

When preparing to present a poster, there are several important questions to ask about the venue (see text box). These questions are largely logistical and need to be answered well before the date of presentation. Posters may remain displayed at times when the speaker is not present. For this reason, the posters must be self explanatory and organized in a way viewers can easily follow the flow of information from beginning to end.

## The Presentation

The author of a poster session generally does not give a prepared speech. Rather, properly designed posters speak for themselves. In fact, the audience may wander in and out, moving singly or in groups throughout the allotted time. Some of the audience may have only cursory interest and read the highlights from a respectful distance and nothing else. Others may want to go into great detail and crowd up next to the poster to read every word. In either case, they proceed at their own pace. Throughout this time, the author is on hand to answer questions, defend the work, or debate issues.

If the venue requires that you give a prepared speech, keep it very short (five to 10 minutes) and allow the remaining time for informal discussion and audience browsing of the posters.

## Design Criteria for Posters

Posters do not bear the same restraints as do slides. They can display a great deal of information and, if necessary, the audience can view the information from up close, allowing them to read finer detail. Design remains important, however, and while simplicity is not quite as necessary as with slides, boldness and color are very useful.

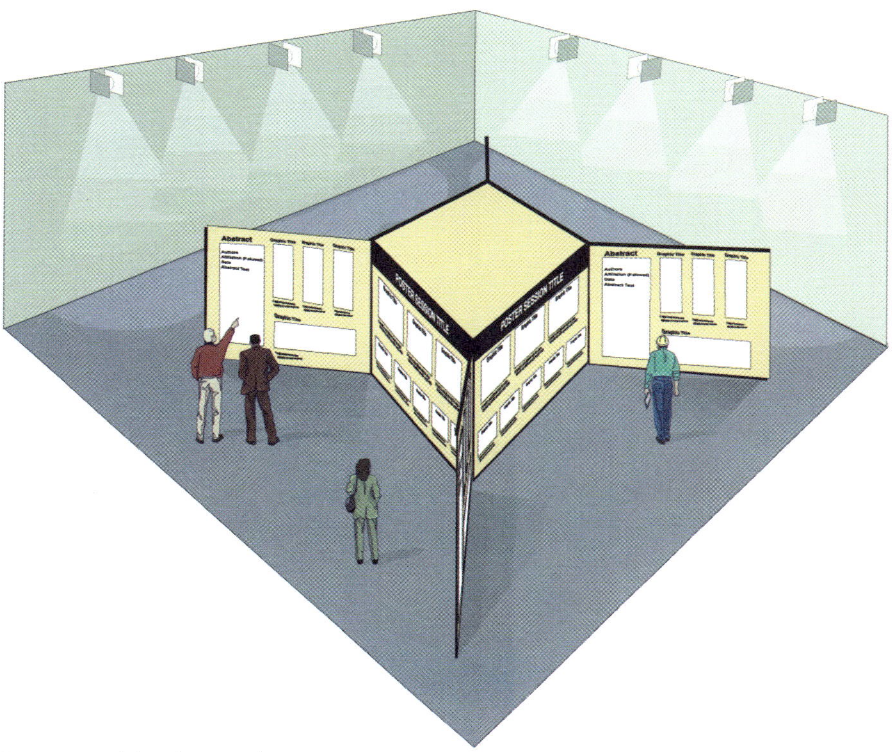


figure 7–1 *A poster session*

## Questions to Ask About the Venue

- *Is a backdrop provided?*
- *What size is the backdrop?*
- *Will the backdrop support velcro, tape, etc.??*
- *Do I need to bring my own tape, pins, or magnets?*
- *Is supplementary lighting necessary?*
- *Is there an electrical outlet?*
- *Can I use supplementary materials (computer, slides, video, table with samples, etc.)?*
- *Are tables provided for supplementary displays?*
- *What are the times for setup, presentation, and tear-down?*

table 7–1 *Preferred text sizes and viewing distances for poster sessions*

| | Size (in.) | Size (pt.) | Viewing Distance (ft.) |
|---|---|---|---|
| Title | 3–4 | 216–288 | 15–20 |
| Graphics Titles | 1 | 72 | 10 |
| Abstract and Conclusion Text | 0.3 | 24 | 6–7 |

Figure 7–2 illustrates a possible layout for a poster. (The space provided in this example is generous. Not all venues offer three side-by-side boards, nor do all posters require that much space.)

Features of a poster include:

- Title
- Author's name and affiliation
- Abstract of the talk
- Annotated displays of technical information
- Conclusions

All of this shoud be presented so that the general flow of information proceeds from left to right, leading from the abstract to the conclusions.

Make it easy for the viewer to follow the development of ideas. Use subtitles, color, and possibly arrows to move from one step in the presentation to the next. Borders or a common background color can group text and illustrations that belong together.

Table 7–1 suggests lettering sizes for titles, captions, and text. These reflect a hierarchy of headings, with the title legible from the greatest distance, subtitles at a lesser distance, and text in the abstract, conclusions, and the data displays at a normal reading distance.

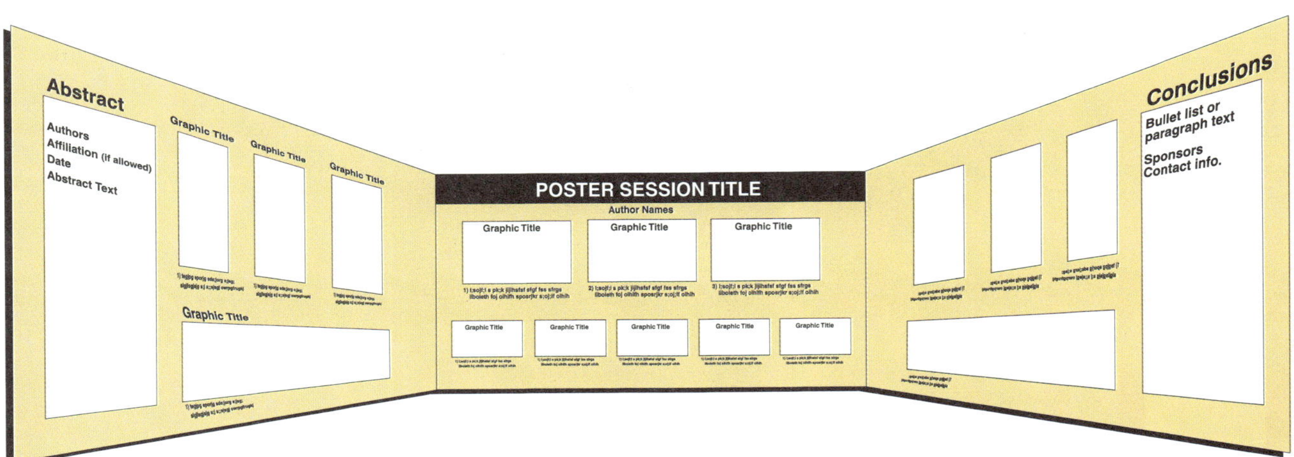


figure 7–2 *A possible arrangement of posters on three boards. Note the flow of information from left to right and the heirarchy of titles.*

# Appendix A

# *Working With a Service Bureau*

## What is a Service Bureau?

Service bureaus are companies that support the digital graphic arts and printing community. They offer expertise in preparing digital files for printing, film imaging, plotting, or other output, and they have the specialized equipment required for these jobs. Many professionals and small companies do not have in-house expertise or equipment for imaging and developing 35-mm slides, producing large-scale plots, or preparing color-separated films for printing. Service bureaus fill these needs.

Types of service bureaus include:

- Pre-press Services: Pre-press service bureaus specialize in producing the color-separated films that are used to produce plates for offset printing. Their expertise is in working with the files of desktop publishing programs, such as PageMaker or QuarkXpress, to output film. Using these films, the printing company prepares plates for printing a job on a press.
- Graphic Arts Services: Graphic arts service bureaus include photo labs, sign shops, slide houses, and color copy shops. These produce a variety of products, including slides, posters, overhead transparencies, color copies, short-run brochures, tradeshow panels, signs, banners, and labels.
- Reprographics Shops: Reprographics service bureaus developed to meet the needs of architects and engineers for blueprints, copying, and other document services. They have further developed into providing color services as well. Typically, reprographics shops focus on plotting and copying, rather than graphic arts, but the lines are beginning to blur as they try to meet all of the needs of their clientele. Most are expert at dealing with the output of files from the CAD environment, whereas many of the graphic arts bureaus have very little experience in outputting these files.

## How to Choose a Service Bureau

Service bureaus often provide critical services in the process of creating and producing presentations in the computer age. Whether your company has an in-house department that takes care of these services or you have to go to an independent bureau, contact and interview a service provider early in your preparation for a presentation. This can help you avoid some serious pitfalls and 11th-hour panic.

Designing beautiful slides or posters is only the first step. Difficulties often arise in the output stage. Service bureaus have the expertise to guide you through the file saving and formatting techniques that will allow them to give you the products you request without expensive, time-consuming, and frustrating mistakes. The time required to develop a good relationship with a good bureau is time well spent.

How do you find a service bureau? The best way is through a referral from a satisfied customer who has used a service bureau for work that is similar to yours. Absent a solid referral, you can start in the Yellow Pages by looking in the "Computer Graphics," "Slides and Transparencies," and "Photo Labs" sections. Select a few, and give them a call.

In your first contact, determine if they do the kind of work that you require. Have them describe their services and send you a brochure and price list. If you are a beginner, pay attention to how well they communicate with you over the phone. If they start asking you questions about what you do and how you will be using their products, take that as a good sign. Communication is going to be the key to working well with the provider you select.

Here are some questions to ask:

- What products do you produce? Tell them what you will require for your presentation, and ask if they can provide those products. For example, if you are giving a poster session, find out if they provide large-scale color plots, if they mount plots on foam core, and if they can laminate the displays.
- What software do you use? What digital file formats must I prepare? No single service bureau supports all of the presentation software that exists in the marketplace. You must find out if they can work with the files that you will create. Can they handle both Macintosh and PC files? Is their version of a particular software the same as yours? A bureau that does not support your software may still be able to suggest ways to save the files in a format that they can use. A bureau that simply says, "No, we don't use that," without offering some effort to help out is probably one to avoid anyway.
- How can I get my files to you? You will have to deliver your digital files to the service bureau before they can start work. This will require that you send the files electronically via e-mail or FTP (file transfer protocol) or that you take them on a removable storage device, such as a diskette or a Syquest, Jaz, or Zip cartridge. Ask for their e-mail and FTP addresses, and ask which removable media they accept. A good service bureau may even offer to connect your storage device to one of their machines to download the files.

Compression utilities, such as Stuff-It™ for the Macintosh® and WinZip® for the PC, allow you to either compress files or spread the data across several linked floppy disks for transport. If you use one of these compression programs to put your files on a diskette, be sure that your service bureau has the software necessary to uncompress the file.

Many files can be sent attached to an e-mail message, but some files are too large to take this route. Some internet service providers (ISP) limit the amount of data that can be attached to an e-mail message, and often this limit is a lot smaller than you might think. Their e-mail servers will simply kick the attached files out and not relay them to their final destination. This can happen even with very large service providers. An FTP site for the direct transfer of files to the service bureau can be more reliable for large files.

- Can I talk to the technicians who will handle my job? All service bureaus will let you talk to a salesman, but you need to also be able to talk to the person who actually handles your job. Do your best to get to know the technician so that you can ask questions and receive tips. Remember, technicians have seen everybody else's mistakes and can keep you from making mistakes of your own. Also, in the deadline-intensive service world, a technician who knows you will be most likely to accept your calls, give you priority, and help you through problems.
- What will it cost? Most of the products of a service bureau are sold by the unit. For example, you will receive a price per slide, a price per square foot of large-format plot, a price per color copy, etc. If you prepare your files correctly and avoid rush charges, you should be able to know in advance what your costs will be. However, other charges may run up the bill. Hourly rates will apply to "file intervention," which is the time required by technicians to alter or fix your files. Rush charges will apply to jobs that must be delivered to you sooner than their normal turnaround time. Your failure to allow enough time to work within a service bureau's normal turnaround time may double the cost of your job. And remember, last-minute jobs are also likely to contain mistakes, and you won't have time to fix them.

## *On Time, Within Budget, No Mistakes*

You've designed and drafted the presentation of a lifetime, following all of the guidelines in this book. You know you are going to knock them dead. You have selected the best available service bureau. Now you have to work with that bureau to make sure that you arrive at your meeting with your presentation in hand and ready to deliver. Follow this sequence of steps for working with that service bureau to ensure that your presentation materials are perfectly produced and in your hands in time for your meeting.

- Check It! Go though all of your pages or slides on screen and look for mistakes. Run the spelling checker! Make sure that text and graphics are properly arranged. Anything you leave undone may show up on your final output. Good bureaus will look for little details and make corrections for you, but you cannot count on them to catch mistakes.
- Save It! Save your final file where you can find it and make sure that you send the final version to the service bureau. Some people incrementally save various files with similar names as they work on a project. This can leave a bewildering number of files to choose from, and occasionally the wrong one is sent to the bureau.
- Collect It! Copy your files to the disk you will use to transport them to the bureau. Include your fonts. Font substitutions will cause your text and graphics to shift, destroying the layout of a slide. Don't take a chance—include your fonts!

### File Intervention

Most service bureaus offer file intervention services. File intervention is just as it sounds– the bureau intervenes in your files on your behalf. When output problems occur, their technicians may be required to make changes to your files to make them output correctly. If they are more skilled than you in setting up a graphic in the best possible way, you may also want them to help you improve your layouts. Remember, however, that file intervention is an added cost, and watch your budget!

- Print It! Make hard copy prints of your file and deliver them to the service technician. The technician can then check your files against the prints, possibly catching errors or recognizing problems. Black and white copies are all that is necessary.
- Submit It! Deliver your files along with a description of the required output, your contact information (including an after-hours number if it is a rush job), the deadline, billing information (include a purchase order number if necessary), and where to deliver the final output. If you are sending your files by e-mail, include all of this information in a message.
- Check it Again! Do not rush off to a presentation without checking the products that the service bureau produces. You don't want to view a poorly imaged slide for the first time when it is projected in front of an audience. Check your work, and then check theirs.

# Appendix B

# *Glossary of Computer-Related Terms*

# *Glossary of Terms*

## *A*

**Adobe Type Manager:** A system-level utility for managing fonts and ensuring that Postscript fonts display properly on screen and print correctly.

**Aliasing:** The effect created when objects or lines are interpreted into dots for a raster or bitmap image, creating stairstepped, jagged edges.

**Annotation:** Text or numerical information that explains elements of an illustration.

**Anti-aliasing:** A process that lessens the stairstepped, jagged appearance of bitmapped images.

**Archive:** A back-up copy of important data saved for later use, often stored on a server or removable media.

**ASCII (American Standard Code for Information Interchange):** The most standard and popular encoding method for converting letters, graphics, and numbers into digital format.

**Aspect Ratio:** The relationship of width to height of an image. The typical aspect ratio of 35-mm slides is 3:2; for overhead transparencies, 4:5; and for computer monitors, 3:4.

**AVI (Audio Video Interleaved):** A file format for digital movies

## *B*

**Backup:** A copy of a digital file stored in a separate location or on alternate media to ensure its safety.

**Banding:** An effect in color gradients, where obvious breaks in the gradient create visible bands of color densities.

**Base:** The framework and other essential elements of an illustration, onto which other information is drafted.

**Bezier Curve:** A tool in illustration and drawing programs for creating and representing complex curves, allowing almost unlimited variations of shape and size.

**Bitmap:** An image formed by a grid of pixels. Each pixel is assigned values for density and or color to create a mosaic that creates the illusion of a continuous tone or black and white image.

**Bleed:** Extending an image beyond the edge of the printable page, with precise trimming of the final product to allow the image or color to extend to the edge of the final trimmed page.

**BMP (Bitmap):** A generic pixelated file format

## *C*

**Camera-ready Art:** Artwork in hard copy form that is ready to be prepared for printing or other production.

**CGM (Computer Graphics Metafile):** A file format supporting both raster and vector information.

**CMYK (Cyan, Magenta, Yellow, blacK):** The process colors used in color printing by most devices. Also known as the "subtractive primaries."

**Color Correction:** Adjusting a scanned image for color and density to optimize it and correct for deficiencies in the image or output device.

**Color Proof:** A print used to approximate what the final printing run will look like.

**Color Separation:** The process of separating an image or entire page into its CMYK components for printing on a press. A separate film is produced for each process color.

**Compression:** Any one of several methods of reducing the file size of digital information for storage or transmission.

## *D*

**DAT (Digital Audio Tape)**: A high-volume storage media, with data stored on a linear high-quality audiotape.

**DOC:** The native file format of Microsoft Word.

**Download:** Receipt of files from another computer, now often accomplished through networks or the Internet.

**DPI (Dots Per Inch):** The number of dots an output device (printer) prints per inch, horizontally and vertically. The higher the DPI, the greater the detail and resolution of the printed image.

**Driver:** A small program that enables a computer to communicate with peripheral devices such as printers, network devices, and removable media drives.

**DXF:** A file format for saving and exchanging information in CAD programs.

## *E*

**E-mail (Electronic Mail):** The transmission of messages and attached files via the Internet or an internal network.

**Electrostatic Printer:** A printing technology that draws toner out of a solution via an electrical charge to create an image on the paper.

**Embellishment:** Graphical elements that enhance the attractiveness, cleverness, or salesmanship of an illustration.

**EPS (Encapsulated PostScript):** A file format generated by many graphics software programs, containing both vector and raster information.

## *F*

**FTP (File Transfer Protocol):** A protocol used to transfer files over a network or the Internet. Effective for transferring large files through a relatively error-free link.

**Font:** A style of typeface with a unique design and weight for display of the alphabet, numerals, character set, and punctuation.

**Formatting:** The preparation of a digital storage media or disk to receive data.

**Four-color Process:** The color representation model used by most digital printers and traditional printing presses, forming the spectrum of colors from combinations of CMYK (representing the additive primary colors Cyan, Magenta, Yellow, and Black )

**Framework:** The basic construction of an illustration, onto which data, interpretation, annotation, and embellishment are added.

*G*

**Gamut:** The total color space or set of colors a given device be it a printer or monitor, can represent.

**Gradation:** A smooth transition between colors or shades.

**GIF (Graphics Interchange Format):** A graphic file format developed by CompuServe that is widely used to exchange picture files over the Internet.

**Grayscale:** A representation of an image in shades of gray.

*H*

**Halftone:** An image represented by patterns of dots of various sizes, creating an illusion of a continuous tonal image. Also applied to color printing, where the process colors (CMYK) are also laid down as a series of dots or screens.

*I*

**Image Resolution:** The density of elements forming an image, measured in dots per inch (DPI), indicating the amount of information an image file contains and how much detail is in the file. The higher the resolution, the higher the quality of the image.

**Imaging:** The process of creating hard copy output, such as slides, overheads, or printed matter, from computer-generated files. Also, the capture of a digital image of photographs or documents by means of a scanner.

**Imaging Area:** the area of a given output that you can actually print. Most printers and output device will not print all the way to the edge and hold out a margin area around a printed sheet.

**Index Color:** A pallette of 256 discrete colors.

*J*

**Jaggies:** The stair-stepped or ragged appearance of curves and lines in bitmapped images.

**Jaz Drive:** A removable hard drive manufactured by Iomega that can hold and transfer 1 to 2 gigabytes of data, depending on the drive and media.

**JPEG (Joint Photographic Experts Group) :** A photographic or raster image file format that allows a high degree of image compression for sharing complex and high quality images.

*K*

**Kerning:** The amount of space between letters, which is adjusted for appearance and readability.

*L*

**Landscape:** Page orientation in which the page is wider than it is tall.

**Leading (pronounced like the element lead):** The spacing between lines of text.

**LCD (Liquid Crystal Display) Projector:** A type of digital projector that uses an LCD panel and projected light to display a digital file.

**Line Screen:** The resolution of a halftone dot pattern expressed in lines per inch.

*M*

**Media:** The substrate or material that is printed on with ink or toner. Also refers to the digital material such as tape or diskette.

**MOV:** A file format for digital movies.

*O*

**Orientation:** The rotation either horizontal (landscape) or vertical (portrait) of a displayed or printed document page.

*P*

**PLT:** Plot file, a file generated from CAD programs containing print instructions for plotting

**PMS (Pantone Matching System):** a system of color standardization and color specification for the graphics industry. Usually specified as a numbered color as in PMS 300. Many of these colors are used as custom spot colors for printing with Pantone mixed ink. These colors can be simulated and approximated with CMYK process colors but can not be matched. Up to 60% of the pantone color chart is outside the printable spectrum of the CMYK process.

**Palette:** A selection of colors for a presentation, or a subset of the colors of the spectrum.

**Peripheral:** A device or piece of hardware than can be added to or connected to a computer.

**Piezoelectric Ink Jet Printer:** Another inkjet technology that uses a piezoelectric actuator to control the flow of ink to the paper for printing.

**Pixel:** The smallest element of a digitized picture on a screen, printed piece, or image file.

**Plotter:** A term used for large format printing devices usually 24" in width or greater. Allows the creation of large displays, posters, and trade show graphics.

**PDF (Portable Document Format):** Adobe Acrobat's native format. Can contain multiple pages with text and graphics. Embeds all information including fonts that are required for the independent printing of a document. A very important format for sharing documents over multiple platforms and computers.

**PostScript:** The programming language that Adobe developed for the description of a page. Basically it translates the page you have created in a computer program or application into the language that your printer understands. A postscript file embeds all of the information and fonts needed to output your file. Using the proper driver you can create files for any output device that can be downloaded to that device for printing.

**Portrait:** Page orientation in which the page is taller than it is wide.

**PPT:** The native file format of PowerPoint, a presentation graphics software.

**Proof:** Hard copy used for review and approval. A relatively accurate representation of the final color as it will look on the final print run.

## Q

**Queue:** the list or line of jobs waiting to be printed on an output device.

## R

**Raster:** An image created by dots or pixels of information. All of the image data is stored as small squares of information like a complex mosaic to give the appearance of a continuous tone photograph or other graphic image.

**RIP ( Raster Image Processor):** A device, including both software and hardware, that translates postscript printer language into the language actually used by a printer to calculate the placement of dots.

**Registration Marks:** Small marks or cross hairs that allow the aligning of multiple layers of film.

**Registration:** The alignment of different films, printing plates, or ink jet heads to produce an accurate printed image.

**RGB (Red Green Blue):** The primary colors used by computer monitors and television screens to create the spectrum of colors.

**RTF: (**Rich Text Format): same text as txt definition

*S*

**Sans Serif Typeface:** A font that lack serifs (small strokes on the ends of the main character stems).

**Saturation:** The purity of a color or the amount of gray in a color. The higher the gray level, the lower the saturation.

**Scaling:** Changing the size of a graphic image.

**Scanner:** A peripheral device that captures photographs or other art and coverts it into a digital file that can be read by and displayed on a computer.

**Serif Typeface:** A font that has serifs (small strokes on the ends of the main character stems).

**SND:** A sound resource file format.

**Storyboard:** A sketched outline of the sequence and content of slides for a presentation.

**SVGA (Super VGA):** A standard for the video displays of computer monitors.

*T*

**TIFF (Tagged Image File Format):** A common and very broadly supported raster file format for scanned images.

**TGA (Targa):** Another common, though less supported, raster image file format.

**Tiling:** Producing large prints by breaking them into several printable sheets and cutting and pasting them together to create a larger sheet.

**Typeface:** A unique design of letters, numerals, character symbols, and punctuation (see font).

**Typography:** The art of setting and arranging type or text for a printed piece.

**TXT:** A text file format that is readable by most word processors.

*U*

**Unzip:** Decompression of a digital file that was compressed using the program PKZip®.

*V*

**Vector:** Files that use a mathematical calculation, rather than pixels, to define the elements of a graphic image, allowing an unlimited scaling of the elements with no loss of sharpness or resolution.

**VGA (Video Graphics Array):** A standard for the video displays of computer monitors.

*W*

**WAV:** Waveform sound file format.

**WMF (Windows MetaFile):** A file format containing raster and vector information used for transferring information between Windows programs.

**WP:** The native file format of WordPerfect®, a word processing software.

**WK:** The native file format of Lotus 1-2-3®, a spreadsheet software.

**WYSIWYG (What You See Is What You Get):** The accurate depiction on a monitor of an image as it will actually print, specifically referring to the postion of text and objects on a page and not necessarily to color.

*X*

**XLS:** The native file format of Excel®, a spreadsheet software.

*Z*

**Zip:** To compress a file for transfer using the program PKZip® on PC's.

**Zip Drive:** A removable disk and drive manufactured by Iomega® that can hold 100 to 250 MB of files for transfer or storage.